Atomic Spectrometric Methods of Analysis

Practical and Technical Guides for Laboratory-based Chemists

Editor-in-chief:
Michael Walker, *Michael Walker Consulting Ltd, UK*

Series editors:
Lucia Burgio, *Victoria and Albert Museum, UK*
Steven Lancaster, *Domino Printing Sciences, UK*
Diane C. Turner, *Anthias Consulting Ltd, UK*

Editorial advisors:
Anthony Gachanja, *Jomo Kenyatta University of Agriculture and Technology, Kenya*
Mathias Schäfer, *University of Cologne, Germany*

For a list of titles in this series see: rsc.li/practguides

How to obtain future titles on publication:
A standing order plan is available for this series. A standing order will bring delivery of each new volume immediately on publication.

For further information please contact:
Book Sales Department, Royal Society of Chemistry, Thomas Graham House, Science Park, Milton Road, Cambridge, CB4 0WF, UK
Telephone: +44 (0)1223 420066, Fax: +44 (0)1223 420247,
Email: booksales@rsc.org
Visit our website at books.rsc.org

Atomic Spectrometric Methods of Analysis

By

Andrew Fisher

University of Plymouth, UK
Email: afisherplymouth@gmail.com

ROYAL SOCIETY
OF **CHEMISTRY**

Practical and Technical Guides for Laboratory-based Chemists No. 1

Paperback ISBN: 978-1-83916-762-1
PDF ISBN: 978-1-83767-276-9
EPUB ISBN: 978-1-83767-277-6
Print ISSN: 2754-7108
Electronic ISSN: 2754-7116

A catalogue record for this book is available from the British Library

The Royal Society of Chemistry is a charity, registered in England and Wales, Number 207890, and a company incorporated in England by Royal Charter (Registered No. RC000524), registered office: Burlington House, Piccadilly, London W1J 0BA, UK, Telephone: +44 (0)20 7437 8656.

For further information see our website at www.rsc.org

For general enquiries, please contact books@rsc.org

For EU product safety enquiries, please email books@rsc.org or contact Royal Society of Chemistry Worldwide (Germany) GmbH, Römischer Hof, Unter den Linden 10, 10117 Berlin.

Printed in the United Kingdom by CPI Group (UK) Ltd, Croydon, CR0 4YY, UK

Preface

This book is aimed at junior technicians and other junior analysts who are just starting their careers or are just moving into a different area of analytical chemistry. In some instances, such people are yet to realise how many errors it is possible to make in the laboratory. The book is filled with possible methods that can be used and/or modified to suit certain sample types. Having personally made many of the errors listed in this book and spent 30 years attempting to prevent students from making the rest, I thought that it would help to present the methods along with some of the most common mistakes so that others know they exist, and can try to avoid them. Sections of hopefully useful hints and tips are therefore provided. Since many of the errors are common to more than one technique, *e.g.* since all of the sample introduction problems are the same for ICP–MS as for ICP–OES, they are only given once. The reader may therefore have to read more than one chapter.

The main thing for an analyst to ask is "where can things go wrong in my analysis?" Once they start thinking like that, making errors is less likely. Have I got a representative sample? Where can I lose my analyte? Where can contamination enter? Have I checked that my tools, *e.g.* balance, pipettes, and analytical instruments are working correctly? Are there interferences? Now that I have my data, am I sure I am doing the calculations correctly? Most importantly, am I obtaining the right answer for my certified reference materials or quality control sample? If not, then at least one thing during the analysis is going wrong.

Practical and Technical Guides for Laboratory-based Chemists No. 1
Atomic Spectrometric Methods of Analysis
By Andrew Fisher
© Andrew Fisher 2025
Published by the Royal Society of Chemistry, www.rsc.org

Given the almost infinite number of ways the human brain can devise of "cocking things up", the text cannot possibly be exhaustive. However, I hope someone somewhere finds my musings helpful.

My thanks go to Dr Mike Foulkes who had the misfortune to read through much of the book, offering useful suggestions in many places. Thanks also to Dr Rob Clough who also contributed a thought or two. Thanks should also go to the Royal Society of Chemistry.

Andrew Fisher

Contents

Practical and Technical Guides for Laboratory-based Chemists No. 1
Atomic Spectrometric Methods of Analysis
By Andrew Fisher
© Andrew Fisher 2025
Published by the Royal Society of Chemistry, www.rsc.org

3 X-ray Fluorescence Spectrometry   43

4 Atomic Absorption Spectrometry (AAS)   73

5 Inductively Coupled Plasma: Optical Emission Spectrometry   104

6 Inductively Coupled Plasma: Mass Spectrometry   128

1 Sample Collection Methods

1.1 Introduction

Sample collection is probably the most important step of the whole analytical process because, if performed incorrectly, the whole of the rest of the analysis is pointless. As an example, if the air quality in the middle of the Atlantic Ocean is to be determined, then attaching a sampling device next to the funnel of a ship would lead to a gross error. This is because particulates and/or gases emitted from the funnel would obviously not be representative of the air as a whole. Obtaining a representative sample (*i.e.* a sample that has the same characteristics of the bulk sample) is the key factor for a successful analysis. Some sample types, *e.g.* air and water, that are fluid, tend to be more homogeneous than others, *e.g.* soils. This should mean that sampling is easier. Despite this, the above example of sampling the air over the Atlantic Ocean indicates that errors can still be made. A second example would be the (relatively) simple task of collecting tap water to ensure that it is safe to drink. If the tap is not left to run for at least a minute prior to collection, then it is probable that water that has been stationary in the pipes for several hours may be collected. This water may have leached metals from the pipes and pipe fittings and will potentially have much higher concentrations of analytes than the water that has been supplied from the reservoir. Analysing the unrepresentative water and then making a complaint to the local water company would be erroneous and embarrassing.

Practical and Technical Guides for Laboratory-based Chemists No. 1
Atomic Spectrometric Methods of Analysis
By Andrew Fisher
© Andrew Fisher 2025
Published by the Royal Society of Chemistry, www.rsc.org

For solid samples, homogeneity can be much more problematic. For instance, when sampling a field of dimensions 100×100 m, then collecting 1 kg of material from one point in the field is exceptionally unlikely to give a reliable picture of the whole field. Certainly, if there is a corner where car batteries have been buried, then sampling a point perhaps 50 m away will not indicate any problem. Similarly, if the sample is collected from the top 10 cm of soil, then a very limited perspective of historical contamination will be obtained. Depending on why the field requires analysis, then this may or may not be a problem. For instance, if potatoes are to be grown, then soil that lies 1 m below the surface that has been contaminated by an ancient Roman smelting site is not likely to be terribly problematic. However, if building work to construct a new housing estate is envisaged for the field, then soil of at least 1 m depth is likely to be disturbed and brought to the surface so that suitable foundations can be built. This then means that the topsoil has become contaminated and may be unfit to grow vegetables and could become a potential health issue if passed from hand to mouth, smaller particulates being inhaled as dust, *etc.*

Another factor to be considered is the regularity of sampling. When analysing trace elements in a river, a single collection will potentially give a snapshot of the water quality at that instant but gives no indication of historical concentrations. Similarly, it will give no indication of water quality further downstream or, if there has been a single source of contamination upstream, where that may be, how much it has been diluted, *etc.* If the river quality is to be monitored regularly or continually, then different or specialised collectors may be required rather than a simple container.

It is therefore necessary to identify what is actually required from the analysis so that an appropriate sampling regime can be formulated. This chapter will attempt to highlight some of the errors and pitfalls of sampling different materials.

1.2   Sampling of Waters

As with any analysis, contamination should be kept to a minimum. Therefore, collection devices should not contain significant concentrations of the analytes under investigation. Sample integrity should be maintained, *i.e.* steps may have to be taken to ensure that the analyte is not lost from the system, *e.g.* by becoming adsorbed to the collection container walls or lost *via* passing through low density polyethylene container walls (in the case of Hg). If a speciation analysis is to be undertaken to determine the overall toxicity potential of an element, then steps to ensure no changes in the oxidation state or to the chemical form in which that element is present should be taken. The majority of samples are stabilised in nitric acid because it does not cause analytes to precipitate. However, the trace metals analyst should recognise that many water chemists will also want to monitor the level of nitrates present. If such a requirement exists, then an alternative stabilisation acid will be necessary, so care should be taken to ensure that the analyte is not precipitated (or co-precipitated). In addition, an oxidising acid such as nitric acid may not be suitable to stabilise waters if a speciation analysis is to be undertaken. For example, a paper that discussed the stabilisation of As species in waters prior to analysis was published by Kumar and Riyazuddin.[1] Similarly, if organic analytical chemists wish to measure organic components, then plastic containers will not be suitable for them. Under such circumstances, two separate samples need to be collected: one in a plastic container (for trace metals) and the other in a glass container.

Collection and/or storage containers for the determination of trace metals should be inert, *i.e.* a pre-cleaned and rinsed container made from plastic. Glass is not suitable because it has active sites on its surface, and these sites can adsorb metal ions. The type of plastic used can also have a large effect. Polyethylene terephthalate (PET) contains significant amounts of Sb, so if this element happens to be one of the analytes, a PET container should not be used. Similarly, polyvinylchloride also contains several organometallic elements and so its use should either be avoided or it should be tested to determine if analytes are leached from it. Most frequently, the containers used are made from high density polyethylene and polypropylene. Different laboratories have different cleaning regimes, but most use a surfactant-based wash, rinsing with ultra-pure water, an acid soak (5–20% nitric acid) for a period of 24 hours followed by a last rinse with ultra-pure water. Lids should be treated in the same way. The containers and lids must then be left to dry – usually not using a paper towel as this can re-introduce contamination.

For river water collection where a limited number of samples intend to be collected, an analyst can enter the water then, facing upstream to ensure that their feet do not disturb sediment that can flow into the container, the analyst can turn the collection container entrance to face downstream so that the water flows around the container and then into it. This protocol minimises the amount of extraneous material, *e.g.* leaves, twigs, and weeds, that is inadvertently collected. It should be noted that although used commonly, this method is not really suitable for the collection of large volumes of numerous samples because of the problem of transporting possibly hundreds of samples each of a litre volume back to a laboratory. It should be noted that the container should be clearly labelled using a permanent marker pen or some other permanent methodology prior to it being used to collect the sample. This is especially true for the "permanent" marker pens that do not work properly on wet surfaces. Ideally, both the lid and the container body should be marked to ensure that accidental swapping of lids and hence cross-contamination and confusion does not occur. It should be noted that the water sample should be filtered prior to stabilisation using acid; otherwise, trace metals will be leached from any particulates present into the water, hence artificially elevating the analyte concentrations present. Depending on the local geology, if such an error occurs, it is very possible to inadvertently elevate the concentrations of Ca, Fe, Mg, Mn, *etc.*

Numerous other commercial water sampling devices have been developed including Kemmerer and Rhizon. A good overview of samplers for assorted water types was presented by the US Geological Survey.[2]

The cost in terms of personnel, transport costs, *etc.* associated with the use of "grab" sampling discussed above can be appreciable. Therefore, automated samplers may also be employed. Examples include the Teledyne-ISCO series of samplers and others available from Sigma, Hach and Sirco. These can collect over 20 samples of volume 500 or 1000 mL and can provide high resolution sampling for several days or possibly weeks. However, they are not well suited for operation in freezing weather or for long duration unattended sampling at remote sites. Other examples exist, including a device developed by Chapin and Todd who used a mini-sipper to collect small volumes (~5 mL) of acid mine water.[3] This device did not use discrete containers. Instead, it collected sample, filtered and acidified it on-line and then inserted an air bubble to separate that sample from the next. Over 250 samples could be collected using the device before samples needed to be transported back to the laboratory. A brief review of some of the automated samplers available was presented by Chapin in 2015.[4]

In general, these types of samplers give a much better view of the state of the water course since they provide samples over lengthy time periods. If samplers are placed at intervals along a river, it may be possible to identify when and approximately where a contamination event occurs. This is far less likely to be possible with "snapshot" samplers, *i.e.* the ones that require an analyst to collect the sample manually.

For water collection at sea, Niskin bottles or Go-Flo bottles arranged in a "rosette" may be lowered over the side of a ship and the devices programmed to open and shut at the desired depth. Once the samples have been collected, they should be filtered and acidified as usual. Storage should normally be in containers that have been cleaned as described above.

For virtually all water sample types, care should be taken to ensure that minimal contamination occurs through sample handling. This is because most water types have high concentrations of "major" analytes, *e.g.* Ca and Mg (plus K, Na, S and others for seawater) but many of the analytes of interest are likely to be present at $\mu g \, L^{-1}$ levels or below. Being at such low concentration, they are at greatest risk of significant contamination. Even seawater that has percentage levels of Na and hundreds of $mg \, L^{-1}$ concentrations of other "major" elements, the other analytes may be present at concentrations significantly below the $\mu g \, L^{-1}$ level. Given the ultra-trace level of many potential analytes, the acid used to stabilise the sample should be of very high purity. Many laboratories that cannot afford the expense of ultra-pure acids resort to sub-boiling distillation of cheaper, poorer grade materials. When filtering the waters, the filters should be pre-cleaned prior to use by filtering some dilute acid through them. These washing can then be discarded and the filters rinsed with ultra-pure water. The containers used to collect the filtered samples should also be pre-cleaned to minimise contamination. Wherever possible, all sample manipulations should take place in either a clean room or a laminar flow hood (or both). Laminar flow hoods and clean rooms are areas where air is passed through HEPA filters which remove airborne particulates such as dust. Both are under positive pressure, *i.e.* the air pressure is slightly higher than that outside. This also enables "dust" to be blown off the analyst before they walk into the clean room and stand next to the laminar flow hood. A laminar flow hood placed in a clean room means that the air present would be doubly filtered. This could be useful since clean rooms vary in quality, with the cleanest being significantly more expensive than the least. Some workplaces may require the analysts to wear disposable suits and head and shoe coverings to minimise contamination further. The clean room often has an ante-room where the analyst enters, walks across a sticky mat to remove contamination from the soles of their shoes and a small bench where they can sit to put on the protective equipment. If a standard operating procedure is available at the workplace, then this should always be adhered to.

1.3 Sampling of Soils and Sediments

Numerous sediment samplers are available and the one to use depends on the amount of sediment required, the depth from which it must be collected, *etc.* The analyst should be clear of the requirements of the analysis to be undertaken. If a very limited study of the topsoil is required, then a garden trowel may be used to simply collect the top 2–5 cm of soil ensuring, as far as possible, that extraneous materials such as leaves and twigs have been removed prior to sampling. The soil collected may be placed in a labelled container to ensure that the place from where it was sampled (a GPS position or "what3words" would identify the position to within a less than a metre), the depth (if necessary) and other useful information such as the date it was sampled and the name of the analyst collecting it, can easily be identified. The sample collection can then be repeated at other sites over the area being studied. Some hand-held corers can only be used in relatively shallow water, whereas others may be lowered to the bottom of the seabed. Some, *e.g.* the hand-held corers, collect only small amounts of sediment, whereas some of the grabs are capable of collecting large amounts. Some are made of plastic whereas other are made of stainless steel. In general, the stainless-steel collectors can be used because the concentrations of analytes in the samples are significantly higher than in waters. It is therefore more difficult to contaminate at a significant level. Several soil corers are available commercially and differ in design, length and diameter. It may even be possible to hammer some rigid plastic piping into the soil to obtain a core. Alternatively, soil augers are also available commercially which, again, have different lengths and diameters. All of these samplers will provide a relatively small sample mass. A sediment grab is capable of a collecting much larger masses of material. This type of sampler has two large bucket-type jaws that can be forced together using a strong spring or, for the largest grabs, hydraulic pressure. The largest ones can collect huge amounts of samples. In areas of large rocks or boulders, the grabs have a tendency to become jammed open and their contents washed away during recovery to the surface. However, the hydraulically operated grabs are the most likely to be capable of recovering cobbles and small boulders than any other system. These huge samplers are most useful when geological exploration is being used to identify sources of economically useful materials.

When an extremely large sample size is taken, *e.g.* using a sediment grab, then the analyst may employ the method of coning and quartering to avoid systematic bias. Coning and quartering may be employed when even several tons of material have been collected and is where the whole sample is placed in an open space forming a mound (a cone). This is then flattened and divided into quarters and two of those quarters that sit opposite each other are discarded. The sample is then collected, mixed and undergoes the cone and quartering process again. The process continues until a manageable sample size is obtained. The method is regarded as being capable of providing a sub-sample that is most representative of the bulk.

The number of samples required to ensure that it is representative of the site should also be considered. An extremely useful guide to the number of samples to be taken for a given area as well as numerous other hints and tips were given in the book by Evans and Foulkes.[5] Also covered in that text are some sample collection protocols, some sample preparation procedures, quality control and the use of reference materials. This book is not a normal textbook and would be an extremely useful read for students, inexperienced professionals and for even experienced analysts who are not used to collecting samples.

If temporal information is required, then a core should be obtained. Here, the older soil will be at the bottom of the core and the most recent at the top. Once collected, the core should be removed from the corer and wrapped in plastic wrap to maintain integrity. The plastic wrap should be labelled with the place of sampling, the top and bottom of the core and other potentially useful information such as date of sampling, the analyst, *etc.* It may then be transported back to the laboratory. Once returned, its treatment will depend on what the requirements of the analysis are. It may be carefully unwrapped, sliced into 1 cm sections representing different time periods, each individual section placed in a labelled container and then dried using one of the techniques described below. In that way, a depth profile analysis may be undertaken. It should be emphasised that each section should be carefully labelled with at least the name of the core from which it was taken and the depth.

The type of sediment or soil sampled may determine the size of the sections taken. Peat can be up to approximately 90% moisture. Therefore, once dried, a huge weight loss will result. This may leave a dry weight of only 0.1 g which is less than that normally taken for an analysis. This is a small but actually quite important point because if the analytical instrumentation available to a laboratory is only a flame atomic absorption or an inductively coupled plasma-optical emission spectrometry instrument, then sensitivity may be insufficient to determine all of the analytes. Alternatively, if a small volume of sample digest is produced, the flame-based instrument may not have enough sample to determine all analytes. It would therefore be advised to use a wide-bore corer so that the 1 cm slice actually has a much larger diameter and, hence, mass.

Once collected, sediment and soil samples should usually be preserved prior to analysis. This is to prevent biological activity transforming some analytes into different forms, *e.g.* bacteria transforming S into hydrogen sulfide. Depending on the amount taken, this may involve simply putting it into a labelled sediment bag, ensuring that the pen used to write on the bag does not become smudged/washed away when water drips out of the bag. The bags may then be oven dried where 110 °C is common for most analytes, but a much lower temperature may be employed for a much longer time if Hg is amongst the analytes to be determined. The drying period depends on the amount of sample to be dried and on how wet it is. Typically, drying 500 g samples of sediments could take at least 24 hours to dry at 110 °C. The analyst should remove the samples from the oven, allow them to cool to room temperature in a desiccator, weigh them and then place them back in the oven for a further period of time, *e.g.* one hour. The process should be repeated until no further loss of weight is obtained in any of the samples. Alternatively, excess water may be drained, the samples frozen and then the material freeze-dried. As the freeze drier starts to operate, the pressure in the sample chamber decreases as it becomes evacuated. The samples can be regarded as dry when the vacuum does not improve any further. The period required for drying in each case would depend on the number and size of the samples. In general, the freeze-drying process is likely to take longer than drying in an oven, but has the advantage of not running the risk of losing volatile analytes such as Hg. The drying processes are discussed at greater length in Chapter 2.

The two drying techniques described above are clearly only suitable for relatively small sample sizes. If a ton of material has been collected, it is wholly unrealistic to expect to be able to use these two methods for drying. It will be necessary to obtain a smaller sample size while ensuring that the subsample obtained is representative of the bulk sample. A sub-sampling technique such as coning and quartering could then be employed. Once the sub-samples of soil or sediment have been dried, they may be stored until further sample manipulation is required, *e.g.* grinding and sieving, acid digestion/ fusion, *etc.*

It should be noted that coning and quartering is not limited to just soils and sediments. Consider, for instance, a large hopper load of animal feed that has been prepared and contains several mineral supplements and other materials of different particle size. It will be erroneous to just take a small sample from the bottom of the hopper because those particles of smaller size are likely to have slipped between larger particles and are present at higher frequency at the bottom compared with the top. Taking such as sub-sample would therefore be unrepresentative of the bulk material. Coning and quartering would help ensure that a more representative sample is obtained.

1.4 Sampling of Air or Airborne Particulates

Airborne particulates are usually collected by using a pump to pass air through a filter at a known volume per unit time and for a known time period. Therefore, the volume of air passed through may easily be calculated. The filters and particulates on them may then be acid extracted or digested and the concentration of metallic contamination determined. It is important to note that not all filter types are metal-free, with glass fibre ones being renown for being "dirty". These can potentially be "cleaned up" prior to use by soaking in acid and then rinsing with ultra-pure water and drying. This though is time-consuming, and hence, many analysts prefer to use cleaner filters, *e.g.* those made of PTFE. Other analysts use a 'belt and braces' approach and employ an acid clean-up of the cleaner filters.

Personal air samplers may also be employed if an occupational hygiene scenario is at play. These samplers may simply be clipped onto the lapel of workers as they go about their business at their workplace. They are demountable and so the filter inside may easily be removed and analysed. As with other filters, if they are touched, the analyst should use gloves or non-metallic forceps.

In both of the above cases, once removed from the holder, filters should be stored in clearly labelled containers that are unlikely to lead to contamination and may be sealed to ensure the filter is not lost. An example could be a Petri dish. Again, the label should be made indelible using either a permanent marker pen or by attaching a sticky label with the writing made using a means that cannot be smudged or erased easily. Both the base and the lid of the dish should be labelled to decrease the possibility of accidentally switching the lids between two different samples.

Gaseous organometallic vapours are rarely determined in air. However, if they are to be determined, then a container such as a Tedlar bag could be employed. Some of these come with metal components, *e.g.* a stainless steel valve. These bags should either be avoided or tests made to ensure that they do not lead to contamination/adsorption of the analytes prior to analysis.

1.5   Sampling of Industrial Materials

The sampling of industrial samples will depend on the industry in-volved, the number of samples prepared per unit time, the standard operating procedures of the facility, *etc.* A factory producing a million individual blood collection tubes every day will not have the capacity to analyse even 1% of them. Instead, they may just collect about 5–20 samples every week and/or when a different batch of raw materials is delivered. If the analysis of a new batch shows no contamination, but analysis of samples from the same batch 3 or 4 weeks later does show evidence of trace elements, then this indicates that some problem has arisen. Further investigation will be required. It is possible that one (or more) of the machines used for the preparation may be starting to breakdown or wear, hence introducing the contamination. Under such circumstances, traceability is imperative, *i.e.* being able to identify which manufacturing machine made those samples. Additional samples may then be taken from that machine and tested further to confirm whether or not that machine needs to be shut down and repaired. Alternatively, contamination of the samples could have arisen through random error during either the sampling or preparation for analysis. This further testing would help clarify this. Analysis of new batches of raw materials would identify if any of these were contaminated at an early stage.

A similar scenario is applicable for other industries that produce samples in such numbers, *e.g.* the pharmaceutical industry. Other industrial manufacturers may produce only a few samples a day. For instance, a steelworks may have several furnaces, each producing different steels. It may be necessary only to collect one sample from each furnace per day. Many tons of steel may be produced from each furnace and so collecting a representative sample is imperative. His-torically, this could involve collection of a tiny sample (often using a type of ladle on a long handle) which is then transported back to the laboratory and analysed using either arc/spark optical emission spectrometry on the solid material directly or acid digested prior to another atomic spectrometric analysis. Clearly, this involves an operator coming into fairly close proximity to extremely high temperatures and is time-consuming. The drive for increased efficiency and cost-effectiveness has driven the requirement for on-line analysis and so these sampling methods are increasingly becoming less necessary.

Numerous other industrial sample types exist and each will have their own sampling regime. It may be necessary to analyse the raw materials in the cases of glasses and ceramics when a new batch arrives. Analysis of the finished products will also usually be necessary for quality control purposes. Again, more than a few samples per day may well not be necessary.

In general, each workplace will have its own sampling regime and standard operating procedure detailing how frequently, how many and how to collect the samples for analysis. A junior analyst should stick rigidly to the SOP unless a lab manager or senior colleague says otherwise.

References

1. A. R. Kumar and P. Riyazuddin, *TrAC, Trends Anal. Chem.*, 2010, **29**, 1212–1223.
2. USGS, Book 9, National Field Manual for the Collection of Water-Quality Data Chapter A2 Selection of Equipment for Water Sampling. Version 3.1, 2014, http://water.usgs.gov/owq/FieldManual/.
3. T. P. Chapin and A. S. Todd, *Sci. Total Environ.*, 2021, **439**, 343–353.
4. T. P. Chapin, *Appl. Geochem.*, 2015, **59**, 118–124.
5. E. H. Evans and M. E. Foulkes, *Analytical Chemistry, A Practical Guide*, Oxford University Press, 2019, ISBN-13:978-0199651719.

2 Sample Preparation Methods

2.1 Introduction

Sample preparation is one of the most important, but often overlooked, factors in a successful analysis. There is no point in undertaking an analysis in which the sample preparation method used does not transform the analytes of interest into a form in which they are readily analysed. For example, if Si is to be determined in a soil sample and a nitric acid digestion is employed, then this will lead to error since the Si-based matrix of the soil does not dissolve. The liquid extracts will give an indication only of the soluble Si rather than the total. However, there are occasions when the "total" of an analyte is not required. For instance, if the concentration that is biologically available to a plant is to be determined, a much milder extraction system, *e.g.* shaking the soil with a dilute EDTA or calcium chloride solution would be necessary. When selecting a sample preparation method, it is necessary to take into account the sample type, the analytes of interest and the method of detection. For instance, there is no point in attempting to determine Cd and Hg in a soil sample using a fusion preparation protocol followed by analysis using inductively coupled plasma–optical emission spectrometry (ICP–OES). The vast majority of soil samples will have low or very low concentrations of Cd and Hg. The Hg and probably the Cd will be lost at the elevated temperatures required for the fusion, and the ICP–OES would not have sufficient sensitivity to determine whatever was left (or indeed, whatever had been there in the first place). In some

Practical and Technical Guides for Laboratory-based Chemists No. 1
Atomic Spectrometric Methods of Analysis
By Andrew Fisher
© Andrew Fisher 2025
Published by the Royal Society of Chemistry, www.rsc.org

circumstances, a workplace may have only one means of detection. Under such circumstances, the analyst should choose their sample preparation methodology very carefully and possibly alter the sample analysis system, *e.g.* by using a preconcentration procedure or perhaps utilising a more efficient sample introduction system, *e.g.* vapour generation instead of the usual nebuliser/spray chamber assembly.

Unfortunately, the sample preparation stage is often the most time-consuming part of the analysis. Therefore, a lot of recent research has focussed on decreasing the time required, simplifying the process, decreasing the number of reagents (and hence potential contamination) and automation.

A possible flow chart of the processes required and possible routes of sample preparation are shown in Figure 2.1. This example is for geological/environmental solid samples. However, it is possible to adapt it to the analysis of other materials. It should be noted though that many industrial-type samples are homogeneous and will not require such lengthy preparation steps. For instance, a steel sample, when analysed for bulk composition, may simply undergo an acid digestion (or be analysed directly in the solid state using XRF or LIBS).

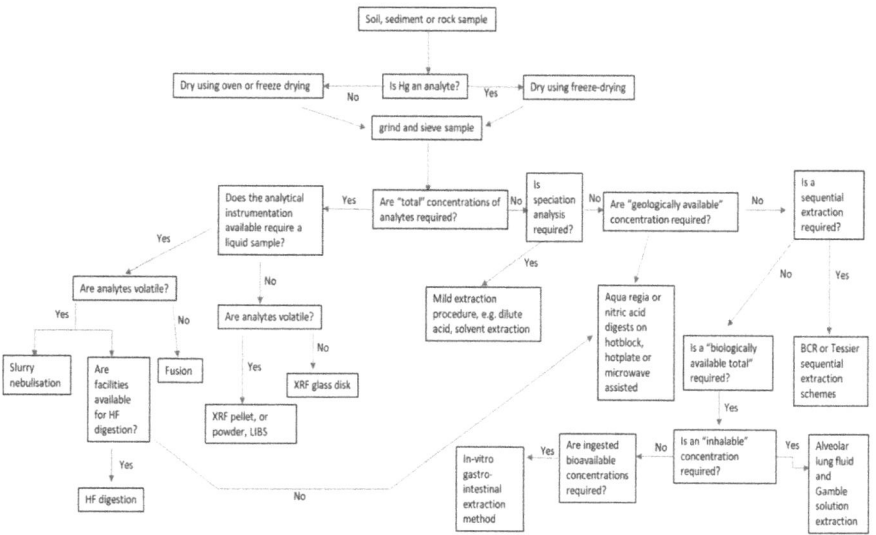

Figure 2.1   A flow diagram of possible steps to take for the preparation of a solid sample.

2.2   Drying and Sieving Samples

Before any of the preparation methods can be undertaken, the sample must be transformed into a state where the procedure can be performed reliably and safely. Therefore, environmental samples such as plant matter, soils, sediments and rocks are usually dried before use. A clear error will be obtained if 1 g of soil is taken for analysis, but rainwater comprises 0.25 g of the sample. It is therefore usual to ensure that all moisture is removed prior to sample preparation. This may be achieved using a drying oven or by freeze drying. Both methods will require the sample to be placed in a container that allows the water to escape. This could be specialist soil/sediment collection bags, plastic bags (**beware these melting at elevated temperatures**), beakers with foil placed over the top, *etc.* In all cases, samples should be clearly labelled to ensure errors do not occur. The drying oven can be set at 105–110 °C, which will obviously dry the sample more quickly than if it is set at 60 °C. However, there is an increased risk of loss of analytes such as Hg. The duration of oven drying will depend on the sample size. Clearly, 1 kg of sample will take much longer to dry than 10 g. However, drying for 24 hours at 105 °C should ensure that the majority of samples weighing less than 500 g are dried to completeness. At 60 °C, the samples may take 3 or 4 days to dry. If the samples are to be dried in beakers, then the likelihood of airborne contamination should be considered. If the oven is old and rusty, then a film or foil over the beaker with holes pierced in it will help minimise contamination. The samples can be assumed to be dry when they lose no further weight. When weighing the samples, they should be removed from the oven and placed in a desiccator for a period of at least 10 minutes before using a balance. Failure to let them cool prior to weighing may lead to errors because thermal eddies could potentially unsettle the balance. Allowing them to cool on a laboratory bench could possibly lead to them regaining some moisture.

The freeze-drying process occurs in a stand-alone machine attached to a vacuum pump. It takes longer but reduces the risk of loss of volatile analytes. The sample should first be frozen. This may be achieved using a simple household freezer or, if one is available and time is pressing, a −70 °C freezer. The time necessary will again be dependent on the sample size and the temperature. It is a good idea to place the freeze drier trays in the freezer too. Since freeze drying occurs typically at −50 °C, if the trays are already very cold, time and energy will be saved once the process starts. **Be careful when handling very cold trays as they may burn the skin or stick to fingers**. If the material is very fine, *i.e.* the particle size is extremely small, there is a chance that, when under vacuum, the particles will be sucked from the container into the freeze drying equipment. This will obviously lead to the loss of sample and further time being lost through the need to clean the freeze drier. Sample loss can be minimised by sealing the containers with laboratory film or aluminium foil that is secured using elastic bands. These "lids" should be pierced with small holes to allow the water vapour to escape. Once frozen the materials can be placed on the trays in the freeze drier and left to dry. The time required for this step can vary from overnight to several days, depending on the sample masses. The samples can be regarded as being dry when the pressure in the freeze drier chamber decreases no further.

After the samples have been dried, they should be sieved to remove other materials, *e.g.* stones, from soils and sediments. Sieves with numerous mesh sizes are available. It is typical to first use a 2 mm sieve to remove larger particles, *e.g.* stones, and then collect the finer fraction. The fine fraction of the materials can then be ground to a very fine powder. This can be achieved using a mortar and pestle or mechanical grinders/mills. Either can be used for soils and sediments, but rocks will need a mill. The material the grinding elements are made from should be taken into account. Agate is often used because it has a high measure of hardness. If the samples are known to contain high concentrations of very hard minerals, then there is a chance that the agate grinding materials may actually end up being ground by the sample and consequently contaminating the sample with high concentrations of Si and other analytes, *e.g.* Ca and Mg. For rock samples or samples known to contain hard minerals, a tungsten carbide swing mill can be used because it is one of the hardest materials known. Before grinding occurs, the origin and nature of the sample should be evaluated. If it is known to be mine spoil from a lead/arsenic mine, then raising dust that is of sufficiently fine particle size to be inhaled is obviously a bad idea. **Gloves and a face mask should be worn and, if other workers are in the vicinity, sample manipulation should occur in a dust cupboard**. Any stray particles of the sample that have escaped onto the bench may then be removed before any further work there commences. If necessary, the ground samples can then be passed through a much smaller mesh, *e.g.* <180 μm or <63 μm. Again, the mesh material should be considered, with brass sieves being less appropriate for trace metal analyses than nylon sieves. It should be noted though that nylon meshes are quite delicate and can easily be distorted if the sample is forced through with a brush. A brush should therefore never be used on the mesh surface itself. Instead, it can be used to disaggregate particles on the rigid sides of the sieve and then the sample is loosely shaken. The grinding and sieving process is often one of the longest processes of the entire analysis. The standard method ISO 11464:2006 would be a good place to start for an analyst wanting to pre-treat environmental samples prior to analysing them.

This chapter will summarise many of the more common sample preparation methods, giving their advantages and drawbacks as well as stating common errors and how to avoid them. The methods discussed here will be those that are used mainly for analytical techniques requiring a liquid sample introduction, so will include different acid dissolution/extraction methods, alkaline extraction, fusions, sintering, slurry preparation, enzymes extractions and dry ashing. Regardless of the method used for sample preparation, it is necessary to have sufficient material available so that replicates can be made of each sample. When weighing the samples prior to preparation, it is necessary to have only approximately the same masses in each sample. For instance, if 0.2500 g is required, then 0.2300–0.2700 g is adequate. There is no need for the analyst to try and use exactly 0.2500 g, as this will be excessively time-consuming. The more the sample is touched, the more likely that contamination will occur. Therefore, ideally, spatulas should not be placed into the sample (and absolutely never into certified materials). Instead, the sample should be tapped out of its container into the preparation vessel. If an excess is added accidentally, then some may be removed and discarded (or used as another replicate) rather than being put back into the storage container. It should be noted that the amount of sample required per replicate to ensure it is representative of the whole sample will depend on the particle size it has been ground to, with smaller particle sizes requiring lower mass.

2.3   Acid Dissolution Methods

The analyst must first ensure that they are absolutely clear what their task is. For instance, there are many approaches to acid dissolution of environmental samples, *e.g.* soils or sediments. Many of these are actually not complete dissolutions but are aggressive leaches. An aggressive leach with nitric acid is sufficiently powerful usually to extract >90% of analytes such as Cd, Cu, Fe, Mn, Pb and Zn, but is often less efficient at extracting analytes such as As, Sn and Ni. For analytes such as these, a more powerful extractant, *e.g.* aqua regia (3 : 1 mixture of hydrochloric and nitric acid) would be more efficient. In both cases, a solid residue will remain after the extraction which mainly comprises aluminosilicates but will also have significant quantities of other analytes such as Sr and Ti that are not completely extracted. For these reasons, it is always necessary to prepare and analyse certified or standard reference materials that are appropriate to both the sample type and the task required. Many certified materials give more than one set of values, *i.e.* they give "total" concentration data as well as concentrations that are "acid extractable". The certificate should specify whether the latter is for nitric acid or aqua regia.

Acid dissolutions/extractions may be done using a hotplate, a hot block or in microwave ovens. In all cases, the vessels in which the digests are to be undertaken and the volumetric flasks used for dilution as well as associated stoppers and lids should be pre-cleaned with acid to ensure contamination from previous samples, detergents and tap water does not occur. Nitric acid with a concentration of 5–10% is suitable for this task with the materials being soaked for several hours. They can then be rinsed with high purity water and inverted to let them dry. The entire inner surface of the vessels should be washed with the acid, and therefore, it is important to avoid airbubbles. When rinsing the acid away from the glassware using ultrapure water, it is necessary to also wash the outside of the vessel. The containers should then be dried. A damp outer surface of a digestion vessel or volumetric flask will prevent the sample being labelled correctly. On glassware, even dried acid residues can lead to indelible pen markings being erased more easily. The acid within the acid bath will slowly increase in concentrations of contaminants (more so if they are not rinsed with pure water before being soaked), to the extent where they may potentially contaminate the glassware rather than clean it. The acid should be changed periodically to ensure this does not occur. The regularity of changing the contents of an acid bath will depend on its frequency of use, the amount of glassware being cleaned and how well it was rinsed before it was immersed.

Numerous methodologies have been developed. This chapter is not designed to give a comprehensive discussion or to list all of them. Instead, it will refer the reader to some standard ISO methods which are an excellent place to start. It will also give a few examples highlighting where errors can be made and ways of preventing them.

A fuller description of methods can be found in a book chapter by Hu.[1]

2.3.1 Hotplate Dissolutions/Extractions

There are many different extractions that may be undertaken on a hotplate. These involve those that use oxidising materials such as nitric acid, possibly with the addition of hydrogen peroxide. This type of extraction is appropriate for the digestion of many biological materials, foodstuffs and as a relatively simple leach of materials such soils and sediments. Other types of hotplate digestions include those that use aqua regia (useful for extractions of soils and sediments, catalysts and dissolution of some metallic samples). Reverse aqua regia (3 : 1 nitric acid : hydrochloric acid) is useful for biological materials where analytes that are more stable in the presence of chloride are to be determined (*e.g.* Sn). Total digestions of rocks, soils and sediments may also be undertaken on a hotplate, but only in specialised facilities that have a fume cupboard suitable for the use of hydrofluoric acid.

A typical hotplate dissolution that can be used for the nitric acid extraction of soils and sediments and for the digestion of biological materials such as plant matter and materials including many types of foodstuffs is given below. It may also be applied to the dissolution of many inorganic salts, alloys and metals. A very similar protocol can be adopted for aqua regia digests. It is worth noting at this point that the individual components of the aqua regia should be added separately into each beaker rather than making a large volume of aqua regia in a 250 mL volumetric flask. This is because mixing them by inversion is potentially hazardous and, once mixed, the solution will produce copious amounts of gas. If left in a stoppered volumetric flask, pressure will build up and the stopper is likely to be forcibly ejected.

(1) Accurately weigh approximately 0.2500 g of sample into a pre-cleaned, dry 50 mL glass beaker.
(2) Add 5 mL of (high purity) concentrated HNO_3 to the sample tube.
(3) Procedural blanks should be prepared in a similar fashion but omitting the sample.
(4) Place the beakers on the hotplate, cover with a clean watch glass and leave to cold digest for at least 1 hour prior to heating. This

is so that easily oxidised material may be oxidised prior to heating to prevent an over vigorous reaction from occurring.
(5) After the 1 hour cold digestion, add a further 1 mL of concentrated HNO_3 to the glass beaker.
(6) Turn on the hotplate and increase the temperature in stages until the contents are gently refluxing at approximately 110–130 °C.
(7) Reflux the sample solutions for at least one hour or until fumes are no longer evolved. If the beakers begin to boil dry more acid can be added.
(8) After cooling to room temperature, quantitatively transfer the digest, with filtering if necessary, to a 25 mL acid washed volumetric flask and dilute to volume with 2% HNO_3.

There are several points that must be made regarding the above protocol.

- The acid used for digestion should be of suitable purity to minimise contamination.
- If more than 0.25 g of sample is used, then more acid will be required for the digestion.
- The 1 h cold pre-digestion is necessary because some samples will contain easily oxidisable material that may froth when heated. This frothing may make it overflow the beaker, hence leading to sample loss. Incremental raising of the temperature once on the hotplate will also help circumvent this problem.
- If aqua regia is used as the digesting medium, a lower temperature (100–110 °C) should be used.
- The samples should not be allowed to boil dry. Covering with a watch glass helps prevent this and also enables the sample to condense and drip back into the sample. This refluxing helps prevent the loss of volatile analytes such as Hg, although extended heating will still lead to losses through the beakers' spout. If dryness looks to be happening, further aliquots of acid must be added.
- The method is particularly useful because 25–30 samples may be prepared simultaneously on a single hotplate. **However, the hotplate is not uniformly hot over its surface. Therefore, some samples will digest more rapidly than others.** It may be necessary to move the beakers around the hotplate surface to help minimise this. Clearly, handling hot beakers containing acid has potential safety implications.

- When preparing the digests for dilution, care should be taken to ensure that sample droplets that have condensed on the watch glass are also washed through into the volumetric flask.
- Many analysis techniques use a nebuliser to introduce the sample. Many nebuliser types are easily blocked by particulate matter. It is therefore often necessary to filter the digests prior to analysis. **When filtering the samples, it is necessary to use an acid resistant filter paper,** *e.g.* **Whatman 541 grade or equivalent.** Otherwise, the filter paper is likely to dissolve too. Care must be taken to ensure quantitative transfer of the sample digests to the volumetric flasks.

As discussed previously, hydrofluoric acid may be used to digest rocks, soils and sediments. Extreme caution should be exercised, with suitable personal protection equipment (gloves, rubber smock, facemask) being worn, the digestions taking place in a suitable facility and in the presence of a hydrofluoric acid-trained first aider. A third person to act as a spotter to identify any droplets lost on work surfaces is also useful. Given the hazardous nature of the digest, the specialist facilities required and the drain on manpower, many institutes avoid this type of digestion unless there is no alternative. An additional disadvantage is that it is of very limited use if Si happens to be on the list of analytes, because it boils away as SiF_4. Glassware dissolves in hydrofluoric acid and should obviously be avoided. Instead, PTFE "bombs" are often used, into which the sample is weighed, the hydrofluoric acid added, usually with an oxidising agent such as nitric acid, the bomb sealed and the mixture allowed to boil in aluminium hot blocks on top of a hot plate for 24 hours. After cooling, the bombs are opened (care required!) and heat is re-applied to boil off the acid mixture. The lids should also be placed upside down on the hotplate to allow any condensed acid to evaporate. Residues can then be taken up in nitric acid or a mixture of acids, prior to dilution to volume ready for analysis. The process is extremely long in that it can take up to 72 hours. As an alternative, instead of evaporating the hydrofluoric acid away, it may be neutralised using boric acid. This addition of an extra chemical may have the consequence of increased contaminant levels. Another potential drawback of the hydrofluoric acid digestion method is that insoluble fluorides may be formed with some elements, *e.g.* La. Therefore, if this is present in the samples at very high concentrations, precipitates may form, potentially trapping other analytes within it.

Other acids may be used for specific dissolutions/extractions. Dilute phosphoric acid (~ 1 mol L^{-1}) is capable of extracting As

efficiently from environmental matrices. This is clearly more environmentally friendly and less of a health and safety risk than using
strong acids. Dilute sulfuric acid is capable of dissolving some steel
samples. However, the acid should be at a concentration of 25–30%,
otherwise dissolution will not occur. If elevated temperature is used
for this dissolution, then extra water will need to be added periodically to ensure the acid concentration does not exceed 30%.

2.3.2 Hot Block Dissolutions/Extractions

A hot block digestion procedure is very similar to that using a hotplate. For these, large glass boiling tubes are inserted into a pro-
grammable aluminium hot block. These blocks come in assorted
sizes with spaces for 12, 20, 36 and 50 samples being the most
common. This method has the advantage of being more uniform
thermally, hence enabling safer and more consistent digestion. In
addition, the tubes' length means that even if the sample at the base
of the tube is 120 °C, the top is not much more than room temperature. There is therefore a much lower chance of the sample
boiling dry. They do however have two drawbacks. One is that the
sample cannot be weighed directly into them. Even if they are placed
in a beaker on the balance, it is impossible to tare it. Instead, the
sample must be weighed into another container (weigh boat or
beaker) and then tipped into the hot block tube. This leads to the
possibility of sample loss during transfer. The weigh boat or beaker
should also be weighed after the sample has been transferred to the
digestion tube. A calculation is then necessary to find the exact
amount transferred (assuming no accidental losses). This process
potentially adds the additional problem of calculation error. Another
potential problem is that if the digestion tubes are not completely dry
when the sample is placed in them, the samples may stick to the
sides. If this does occur, it can be remedied by using the digesting
acid to wash it into the base of the tube. If small amounts of sample
remain stuck on the sides of the tube, then, it may be washed off as
the acid refluxes. This is less likely to happen the further up the tube
though.

Once digestion is complete, the tubes may be left in the hot block or
removed and placed in a tube rack or holder. If left in the block,
cooling relies on the rate that the block cools and is therefore a
lengthy process. However, if the tubes are removed from the hot block
and placed in racks they can cool more rapidly. Removal of tubes from
the hotblock while hot should be performed with care to prevent

burns. Similarly, there should be space in the fume cupboard for the storage racks so that noxious fumes are not allowed to escape into the open laboratory. As with hot plate digestions, filtration may be necessary.

2.3.3 Microwave Dissolution/Extractions

If a microwave digestion is to be undertaken, then a pre-digestion is also recommended. However, the actual digestion time is greatly decreased, since with microwave assistance, the time required is only 10–30 minutes. In addition, the temperature and pressure inside the vessels may be monitored and adjusted. This increases the efficiency of digestion. The closed vessels also prevent the losses of volatile analytes and greatly diminish the possibility of airborne contamination. Another advantage is that the carousel of digestion bombs may be 30 or more, enabling multiple digestions to be performed simultaneously. The scientific microwave digestion systems also have a built-in safety mechanism whereby there is a rupture disc that is designed to give way if the internal pressure exceeds a certain limit. The fumes are then directed safely into a fume hood.

Several ISO standard methods exist for the microwave dissolution of materials. Included in this number is ISO 12914:2012, which describes an aqua regia microwave-assisted extraction of soils and soil materials, which is applicable to all soil types and may be adapted for other similar materials such as sediments. If total element concentrations in such materials are required, ISO 14869-3:2001 (the nitric, hydrochloric and hydrofluoric acid digestion of soils) is available. These methods can be adapted to numerous other sample types. In addition, the manufacturers of microwave digestion systems also produce many applications notes. It is therefore worth examining their websites since they can be useful sources of information. As an example, one microwave manufacturer's website[2] gives procedures for a large range of sample types, providing the necessary acid(s), temperatures, times, pressure, whether stirring is required, *etc.* Although the exact conditions required may vary slightly between different microwave systems, it is still a good place to start.

2.3.4 Sulfuric Acid Wet Ash Process

Some samples, *e.g.* plastics, rubbers and materials that are extremely high in fat, such as full fat beef or lamb, oily fish such as salmon or mackerel, *etc.* are very difficult to digest completely using nitric acid.

This is because the fat largely remains undigested and simply floats on the surface of the digest. There is also the potential for the fats to react with the nitric acid to form explosive compounds. In the case of plastics, many are inert and are unaffected by a simple nitric acid attack. A sulfuric acid wet ash is a very convenient way of dissolving such samples. A possible protocol for this method is:

Accurately weigh approximately 0.2500 to 0.5000 g of sample directly into a pre-cleaned, dry, long-necked reflux-type hot block digestion tube.

Add 5 to 7 mL of cold concentrated sulfuric acid to each digestion tube and place in a hot block digestion unit.

Set the hot block digestion unit temperature control to 250 °C, switch the unit on, allow the hot block and digestion tube(s) to come up to temperature and continue heating at 250 °C for 2 hours. The sample will appear as a black tar-like material.

Add concentrated nitric acid dropwise from a Pasteur or dropping pipette to each of the digestion tubes. **Caution! This reaction is vigorous and clouds of dense yellow/brown fumes are evolved**. Add only one or two drops at a time. Take care when adding nitric acid drops to other sample tubes as the fumes from a previous one may burn hands.

Wait until the fumes evolved have subsided before adding another drop or two of concentrated nitric acid to each tube. Add sufficient concentrated nitric acid until the digest turns from a black sludge to a clear yellow/brown colour. This process can take over an hour (depending on sample type and mass taken).

Allow the digests to cool to room temperature. Quantitatively transfer each digest into a separate pre-cleaned volumetric flask and dilute with dilute nitric acid or a mixture of nitric and hydrochloric acids. **Caution – dilution of sulfuric acid is extremely exothermic and hence the digests will become very hot again**. Dilute to volume, wait for the digests to cool and then add sufficient diluent to reach the desired volume again.

Although slightly hazardous, the method is extremely efficient at dissolving plastics and fatty materials. There are a few notes to be made on the above method.

- The method is long, easily lasting 3 or 4 hours. The analyst must be patient because there are no short cuts.
- The nitric acid **must** be added a drop at a time. If a small squirt is added, an extremely vigorous reaction may occur and the sample may exit the top of the tube. Alternatively, if too much is added

over too short a time period, the sample cools to significantly below 250 °C and digestion may not occur so efficiently, thereby increasing digestion time.

- When adding the nitric acid dropwise, care should be taken so that exposed skin is not placed over the top of other digestion tubes.
- The tops of the tubes will be hot, so care should be exercised when touching them.

As discussed previously, the method has numerous advantages over simple nitric acid digestions, including the capability of destroying fatty material and polymers more efficiently and safely. Another advantage is that if Hg determination is required, the tubes are sufficiently long so as to prevent it from boiling away. This is of particular use for the analysis of food samples. The method may be adapted to the analysis of paints, waxes, oils and tars.

The disadvantages of the method centre around sulfuric acid. It is not the cleanest of acids, *i.e.* it often presents relatively high concentrations of contaminants. The very high purity sulfuric acid available commercially is extremely expensive and may not be available for many labs for routine use. This can be overcome to considerable effect by using sub-boiling distillation to purify the cheaper, more contaminated, material. Another potential problem is that some instrumental techniques, *e.g.* ICP–MS usually utilise nickel-based sampler cones. These can be badly affected by the presence of high concentration of sulfuric acid whereby the orifice can enlarge significantly, leading to a failure of the vacuum system. This problem can be overcome in one of two ways. The sulfuric acid can be evaporated almost to dryness in the hot block prior to dilution. However, since the boiling point of sulfuric acid is over 300 °C, this can be time-consuming. The alternative method is to exchange the nickel cones with platinum ones. These are much more resilient to sulfuric acid but are obviously much more expensive.

If the sulfuric acid remains in the diluted digests, then it is worth matching standards with the concentration of it to that in the samples. This is because it is very viscous and will have significantly different nebulisation properties into instruments such as flame atomic absorption, ICP–OES and ICP–MS when compared with standards prepared in dilute nitric acid.

A further problem with introducing sulfuric acid to ICP–MS instruments is that numerous sulfur-based polyatomic ions will be formed, *e.g.* S_2^+ or SO_2^+ (which could interfere during the determination of Cu or Zn over the *m/z* range of 64–68).

2.4 Fusions

Fusions can be very useful for complete sample destruction for materials that are not dissolved easily using acids. Among the most common sample types analysed are geological materials such as rocks, tephra, soils and sediments, although numerous other sample types, *e.g.* ceramics, metal oxides and glasses, may also be analysed. One significant advantage over dissolutions using hydrofluoric acid is that the fusions do not lose Si. Therefore, a fusion is a better option if Si is one of the analytes. The sample type will determine which flux is to be used and the temperature at which fusion should be undertaken. Various fluxes and their applications are shown in Table 2.1

The fusion methods do have several disadvantages. They are very time consuming and labour intensive. They also lead to the prepared samples having a very high dissolved solid loading meaning that they require further dilution prior to introduction to many atomic spectrometric techniques. This is especially so for plasma-based instruments, *e.g.* ICP–OES and ICP–MS that will otherwise have the injector of the torch blocked by the desolvated sample. Other drawbacks associated with the very high salt content include memory effects (*i.e.* long wash out periods between samples). The necessity for very high dilution factors can lead to some analytes being diluted to below the limit of detection of the instrumental measurement technique. The elevated temperature of sample preparation can also lead to loss of volatile analytes. This is therefore an important factor to take into consideration when planning the analysis. Some prior knowledge of environmental samples can also be very useful. For instance, if the sample is a mine spoil high in platinum poisons such as As, S or P or contains high concentrations of analytes that alloy with Pt, *e.g.* Cu, severe damage to expensive crucibles can occur. Similarly, if a fusion using sodium carbonate is to be undertaken, but none is available,

Table 2.1 Different fusion methods.

Flux	Crucible type	Temperature/°C	Sample type
$K_2S_2O_7$	Pt	500	Nb_2O_5, Ta_2O_5, TiO_2
Li borates	Pt + Au, graphite	1000–1200	$Al_2O_3, SiO_2,$
NaOH or KOH	Ag or Ni	750	Glass, porcelain
$KOH:KNO_3$ (7:1)	Ag or Ni	650	Ruthenium, chromite
$NaCO_3$	Pt or Zr	1000	Minerals, silicates
Na_2O_2	Fe, Ni or Zr	700	Rutile, ilmenite, refractories

lithium carbonate may be used. However, unlike sodium carbonate, the lithium analogue is corrosive to platinum crucibles. A different crucible material would therefore be required.

Despite the numerous drawbacks, fusions can be invaluable for the preparation of these refractory sample types since they avoid the use of hazardous hydrofluoric acid. A typical methodology is outlined below:

Pre-clean graphite crucibles by placing some lithium metaborate/ lithium tetraborate flux into them and fusing them at ~1000 °C in a muffle furnace for 10 minutes. The molten glass can be discarded.

Once cooled, the graphite crucibles can have a sample (~0.2500 g) weighed into them and flux (~1 g) introduced. The sample and flux can then be mixed using a plastic spatula, making sure that the graphite is not dislodged. They are then placed in a muffle furnace at fused at ~1000 °C for at least 15 minutes.

A crucible is removed using long tongs and the contents are tipped into a beaker containing ~40 mL of dilute (~10%) nitric acid.

The crucible can then be placed on a heat resistant mat to cool.

The beaker containing the sample can then be placed on a magnetic stirrer, a stirrer bar added and then mixed until complete dissolution occurs. Once dissolved, the sample can be transferred quantitatively to 100 mL volumetric flasks and diluted to volume. Further dilution can then be made, as necessary.

There are several points to be noted regarding the above method:

- Fusions do not have to be undertaken in muffle furnaces, although they do have the advantage of being capable of preparing several samples simultaneously and being more capable of controlling the temperature used. Instead, they can be performed using flames, *e.g.* from a Meker burner – although not with a graphite crucible.
- It is impossible to label a crucible since any markings will burn off in the muffle furnace. Therefore, a clear note of where each sample is in the muffle furnace should be made in a laboratory notebook.
- For safety sake, fusions should be undertaken ideally when nobody else is close to the analyst, since exceptionally hot material will be transferred from the furnace to the dissolving acid in a fume cupboard.
- The graphite crucibles described above oxidise relatively rapidly and may have a lifetime of only 10–12 firings. This number will be dependent on the temperature used and how long they are

exposed to air when the muffle furnace door is opened at elevated temperature. It should be noted that even the metallic crucibles can become brittle and lose material after prolonged use (and potentially contaminate the sample – consider what the analytes are before choosing the crucible type). Glassy carbon crucibles are chosen by some laboratories as a compromise between cheap but easily oxidised graphite ones and more expensive metallic ones. They also have the advantage of having very low metallic contaminant levels.

• Since only one crucible can be removed from the furnace at any one time, it is necessary to shut the door as quickly as possible so that the others do not cool significantly. This will also minimise damage to the crucibles.

• When tipping the fused sample into the dilute acid, there can sometimes be small droplets of it that remain in the crucible no matter how much it is agitated. This can be a source of analyte loss and potential inaccuracy. Depending on the application, it may be possible to use an intrinsic internal standard to correct for this. For example, when analysing rock samples, this can be achieved by doing a replicate analysis of the same sample and then normalising all analytes to perhaps the Si concentration in the replicate with the highest concentration.

• When a crucible is placed on a heat proof mat to cool, it can occasionally have air trapped underneath it, which heats up and expands. The crucible then starts moving across the surface and can cause damage if it leaves the heatproof mat and lands on a fume cupboard surface. This phenomenon can be partially overcome by forming a pyramid of crucibles, which allows air to pass around and under each of the crucibles unobstructed.

• When using the magnetic stirrer to help dissolve the sample, ensure that the magnetic stirrer bar does not spin too fast or jump about. This prevents splashing and hence analyte loss and/or potential contamination of nearby samples.

2.5   Sintering

Sintering is most frequently used for the preparation of geological materials. It is similar to fusions and uses many of the same materials but uses a much lower temperature. Instead of ∼1000 °C, sintering typically uses 400–500 °C. As well as using similar materials, it also suffers the same drawbacks and advantages. A paper by Bokhari and

Meisel discussed the use of varying amounts of sodium peroxide during the sintering at 480 °C of 100 mg of rock samples.[3] The cool sinter is then normally dissolved in either water or dilute acid prior to the analysis. The acid may be necessary to dissolve any insoluble material.

2.6 Dry Ashing

Dry ashing is particularly useful when samples high in organic matter, *e.g.* plant material, animal tissues, *etc.* are to be analysed. Sample is placed in a muffle furnace in a container, *e.g.* a ceramic crucible and heated to perhaps 500 °C reducing organic matter to ashes. After cooling, these may then be dissolved (or extracted) in acid prior to analysis. Consideration of the analytes and of the crucible material should be made to minimise contamination. Sometimes, an ashing aid such as magnesium nitrate solution can be added.

There are several advantages of dry ashing. These are:

- More than one sample can be placed in a muffle furnace at the same time, thereby increasing sample throughput.
- Large masses of organic matter can be used and then the ashes can be taken up in relatively low volumes of acid. A sort of pre-concentration is therefore established, improving the limits of detection.
- The procedure is relatively safe.
- Very few reagents are used. Therefore, contamination is minimised.

The disadvantages of dry ashing are:

- It is a relatively slow and labour-intensive process because it can take several hours for ashing to be complete.
- There is a chance that volatile elements may be lost during sample preparation. This may largely be overcome by the use of an ashing aid such as magnesium nitrate. A 50% solution of this is placed over the sample and dried prior to ashing. During ashing, magnesium nitrate decomposes forming magnesium oxide. The release of the oxygen aids the ashing process and the residual magnesium oxide helps trap analytes inside its matrix, preventing their loss (a similar process to matrix modification during electrothermal atomisation atomic absorption spectrometry).

- Muffle furnaces, especially when older, may drop extraneous materials into the sample, hence contaminating them.
- When the muffle furnace door is opened, air currents can cause low density ashes to be blown out of their containers leading to analyte loss and/or cross contamination of other samples. This can be prevented by placing a lid on the crucibles. This helps prevent sample/analyte loss.

In common with the fusions, it is not possible to label samples and hence a note of where they are placed in the muffle furnace is required to prevent confusion.

If magnesium nitrate is used as an ashing aid, a similar concentration should be added to all standards and blanks in an attempt to matrix match. The viscosity of all samples and hence nebulisation characteristics would therefore be similar.

2.7 Alkaline Extraction

Alkaline extraction is a technique often used for when halides are to be determined, usually in biological materials. This is necessary because acid dissolution is likely to lead to losses of the halides, which may be boiled off as HBr, HI, *etc.* Typically, tetramethylammonium hydroxide solution is used to extract the material at elevated temperature (~80 °C). The procedure is similar to a hotplate acid digestion and similar quantities of sample and volumes of reagent are required. Since the sample is unlikely to dissolve completely, filtration is usually required. Matrix matched standards are again recommended to prevent differences in nebulisation efficiency.

2.8 Slurry Preparation

Some samples that are not easily acid digested or extracted may be introduced to the analytical instruments as a slurry. This is where the particles are ground so that they are as fine as possible so that, when nebulised into an instrument, they behave like a solvent droplet. The fundamental parameters necessary for successful analysis using inductively coupled plasma spectrometry were discussed in a paper by Goodall *et al.*[4] The paper is very old but gives an excellent description and would be a good starting point for anyone new to the area. Basically, the particle size necessary is dependent on the density of the material, with samples of density 1 g cm^{-3} requiring the particle size to be <2.9 μm. For samples with a density of 7 g cm^{-3}, the particle size should be reduced further to a maximum of 1.5 μm. The particles should be prevented from forming loose agglomerates as these too would be discriminated against by a nebuliser/spray chamber sample introduction system. Prevention of agglomeration is usually achieved using a suitable dispersant. Another excellent, but old reference that discusses numerous aspects of slurry nebulisation into plasmas was presented by Ebdon *et al.*[5] It should be noted that slurries may also be introduced to flame and electrothermal atomic absorption spectrometers (ETAAS). The particle size is not so critical for these as the nebuliser/spray chamber assembly for the flame instrument is less discriminating than ICP-based instruments. The main problem with the ETAAS analysis is that if an autosampler is used, the slurry particles will slowly sink to the bottom of the sample cup unless some sort of agitation is applied, *e.g.* a sonic probe. For ETAAS, it may possibly be easier to simply use a manual sample introduction using a pipette, although, this will also lead to the well-known problem of poorer precision. In both modes of AAS, the particles should still be sufficiently small so that a representative aliquot of the sample is analysed.

A typical method is outlined below. It should be noted that the dispersant used and the grinding medium will depend on the sample type and on the analytes to be determined.

Sample, 0.2500 g, is weighed into a pre-cleaned 30 mL capacity screw cap polyethylene bottle. Dispersant (10 mL) (see below) and zirconia beads (1 mm diameter, 5 g) are added.

The bottle is then tightly capped and the bottles placed on a mechanical shaker for 2 hours.

Make sure the shaker is balanced, *i.e.* place samples and blanks diametrically opposite each other. Failing to do this will result in the shaker travelling across the surface it is standing on.

After shaking, the solution is transferred quantitatively through a clean Buchner funnel (no filter paper) into a pre-cleaned, labelled volumetric flask and diluted to volume with more dispersant. After mixing by inversion, the sample may be analysed directly against aqueous standards prepared in the same dispersant.

If analysis is not immediate, the particles will slowly sink to the bottom of the flask. It is therefore necessary to re-suspend them by inversion and/or by magnetic stirring, although the latter may not mix the sample in the neck of a volumetric flask adequately.

Zirconia beads are ideal for the grinding of most materials because they are hard (8.5 on the Mohs scale, where talc is 1 and diamond is 10).

This means that the large majority of samples are softer and can therefore be ground. It should be noted though that a relatively high level of Zr and also Hf will enter the samples because the beads grind each other to some extent (the level of which will be determined by the length of grinding). If either of these elements are on the list of analytes, a different grinding material, *e.g.* agate may be used. A microniser is also a useful method of particle size reduction. If harder materials are to be ground, then a tungsten carbide swing mill can be employed.

There are several things to note in the above procedure:

- Numerous dispersants can be used. Examples include 0.1% Triton X-100 or Aerosol-OT for organic materials such as plants, animal tissues, coal, *etc.* Sodium hexametaphosphate or sodium pyrophosphate (0.01–0.1%) may be used for inorganic-based samples, *e.g.* ceramics, soils, *etc.*
- If extended grinding times are used, the zirconia beads start to grind the plastic bottle, and small plastic flakes will enter the samples.
- Some sample types, *e.g.* some types of plant material, are "spongey" or "fibrous". These may not grind easily as they absorb the force of the zirconia beads. The grinding process may be assisted by immersing the sample in a small volume of liquid nitrogen. This makes them more brittle and so, as the nitrogen boils away, they can be ground using a mortar and pestle (**Caution! Liquid nitrogen can cause severe burns**). This will diminish particle size significantly and is a good method of pre-treating them prior to slurry formation.

A recent review of slurry preparation and analysis using spectrometric methods was presented by Cerquiera *et al.*[6] The review contained 115 references and covered slurry introduction to various instrumental analysis techniques and gave a review of some of the more recent applications.

2.9 Sample Preparation for Speciation Analysis

The topic of speciation is covered in another publication of this series. However, a brief mention here is necessary for completeness. Speciation analysis is especially useful for the study of toxicity for environmental and food matrices. The overall toxicity of an analyte is not dependent on its total concentration, but instead is dependent on the chemical form in which it is present. An example is for As which is generally regarded as being toxic. However, seaweeds contain huge concentrations of it (40–100 mg kg^{-1}) which is well over the recommended limit for human consumption. However, it is present almost entirely as harmless arsenosugars. Hence, As is effectively bound up and is not capable of acting as a poison. Similarly, in fish tissues, it is present mainly as the harmless compound arsenobetaine, with lesser amounts of the more toxic species such as monomethylarsonic acid, dimethylarsinic acid, arsenate and arenite. The fundamental requirement for any speciation analysis is to extract as much of each chemical species as possible without altering their chemical state, *i.e.* to ensure that a non-toxic species in converted to a toxic one or *vice versa*. In general, much more gentle extraction methods are required. When developing a new extraction method, it is important to test it on pure standards of individual species to ensure that no alteration of the species occurs. A good recent review of extraction methods for speciation analysis was published by Martins Viana *et al.*[7] Although the review concentrates mainly on liquid samples, there is a section that discusses the extraction of analytes from solid environmental samples. Also discussed were sample collection, preservation and assorted extraction techniques from liquid samples. There are numerous methods suitable for the extraction of solid samples, but these are often element specific. For instance, dilute phosphoric acid (1 mol L^{-1}) may be used to extract As species.

Enzymes may be used to extract food-based samples, although the enzyme used will be dependent on the sample type. For instance, Trypsin or a protease is most suitable for animal/fish tissues, whereas a cellulase would be more efficient at extracting plant materials. Each enzyme will require its own optimal pH value and buffer. Sample should usually be ground in the presence of buffered enzyme in something like a Potter homogeniser to obtain maximal efficiency. Care should be taken here because if the plunger is depressed too quickly, the sample is likely to squirt out of the homogeniser and go up the analyst's sleeve. An alternative is extraction using an organic solvent. Methanol is often the preferred solvent and can be used either with assistance from ultrasonic agitation or without. Depending on the analytes, some extractions may use acetic acid (plus tropolone), *e.g.* for organotin speciation. Tin speciation is known to be particularly problematic with some species (especially the phenylated ones) being known to exchange groups with other tin species.

For the most recent methods for speciation analysis, the reader is referred to the Atomic Spectrometry Update published in *The Journal of Analytical Atomic Spectrometry*.[8] This review is published annually and is an excellent place to keep abreast of the most recent research.

2.10   Analyte Preconcentration Techniques

Many analytical techniques do not have sufficient sensitivity to detect analytes directly in some sample types. Therefore, unless significant investment is made on newer, more sensitive techniques (plus the infrastructure required to house them!), some form of analyte preconcentration is required. There has been a huge number of papers describing methods of analyte preconcentration published in recent years. Most preconcentration techniques have been developed for the analysis of liquid samples. However, they may be applied to solid materials if an extract is undertaken followed by buffering of the extract and then the method used will depend on the sample type. If a non-saline water is to be analysed for toxic analytes using flame AAS, a simple method of preconcentration would be to obtain a litre of sample and boil it so that the volume reduces to 100 mL. A 10-fold preconcentration would be achieved. Such a method would be unsuitable for seawater because of the high concentration of salts.

This section would be vast if the detail of each method was discussed. Instead, the theory of each will not be discussed. Instead, it will concentrate on the relative advantages and disadvantages as well as potential errors of each class of method. Some of the classical methods of preconcentration, *e.g.* coprecipitation and liquid/liquid extraction, often require large volumes of sample. It should be noted that transporting high volumes of samples back to the laboratory can be problematic. For instance, collecting and transporting 1 L of each of 100 samples from a river bank to a laboratory would involve using a vehicle capable of driving next do a river bank without becoming bogged down. Alternatively, it could park as close as possible and the analysts carry the samples back to the vehicle. This could be time-consuming and energy-sapping. Many of the more common methods therefore require smaller, more easily transportable volumes of sample.

There are some reviews already in the literature that could be of use to the reader. Some are element specific, *e.g.* a review of preconcentration methods for Cd,[9] for Pb,[10] for Cr[11] and for rare earth elements.[12] Although these reviews are old, they can still give the reader an insight into the methodology available. Other reviews are more specific to the method of preconcentration. Examples of these will be given in the relevant sub-sections below.

2.10.1 Coprecipitation

Coprecipitation is a classical wet chemical method of analyte pre-concentration/matrix removal. There are many types of coprecipitation agents. However, one of the most common is magnesium. If a soluble magnesium salt is added to a liquid sample and then the pH is raised through the addition of a base, a precipitate of magnesium hydroxide will collect the analytes. The mechanism could be through adsorption or occlusion. The precipitate may then be separated from the water, dissolved in a small volume of acid and diluted to a known volume with water. If 100 mL of sample is used and the precipitate formed from it is dissolved in 1 mL of acid and diluted to 10 mL, a 10-fold preconcentration occurs. It is possible to obtain much higher preconcentration factors though (>100). If the sample is very high in magnesium naturally, *e.g.* seawater, extra need not be added. Other precipitants include iron III salts, manganese dioxide and lanthanum salts. The process is clearly labour intensive but can be relatively rapid. If the detection method is XRF, then the precipitate need not be dissolved again. Instead, the solid material can be dried and then analysed directly. This would provide a much larger preconcentration factor.

The advantages of the method include its relative ease, the low number of reagents required and the large preconcentration factors obtained. A disadvantage is the high concentration of salt arising from the precipitating material which can potentially be problematic for instruments that use a nebuliser as a sample introduction method. Another potential problem is the incomplete retention of the analyte. The analyst should validate the method using either certified materials of the same matrix or through spike/recovery experiments.

2.10.2 Liquid–Liquid Extraction

This is another classical extraction method that often requires a large volume of sample. Typically, a known volume of sample is buffered to an appropriate pH (often, but not always pH 5.5–6), transferred to a separating funnel and then a small volume of an organic chelating agent in an organic solvent is added. Once stopped, the sample is then shaken releasing the pressure occasionally by either removing the stopper or inverting the funnel and opening the tap. If the tap is opened, it is necessary to ensure that no sample is ejected through the spout. The funnel is then allowed to stand enabling the solvent and sample to separate (the analysts should ensure that they know which layer is which). Once separated the solvent layer is removed and saved. This can be problematic. A Pasteur pipette may be used or the bottom layer run through the tap and collected. Either method runs the risk of accidentally collecting a small volume of the other layer. It is therefore advisable to deliberately leave a small amount of the organic solvent with the sample. A second aliquot of the chelating agent in solvent is then added and the process is repeated. A second repeat extraction is then made and all of the extracts combined. The three extractions are required to make a compromise between obtaining maximal extraction efficiency whilst minimising the time required. If the efficiency of extraction is only 66%, then the first extraction will take 66%, the second 66% of the remaining 34%, *etc.* The overall extraction efficiency would therefore be $66 + 22.43 + 7.63 = 96\%$. With a starting volume of 100 mL of sample and extraction volumes of 5 mL, an overall preconcentration factor of 6.67 is obtained. A higher preconcentration may be obtained by back-extracting in the same way into dilute acid (*e.g.* 3×2 mL of 10% nitric acid). In this instance, a preconcentration factor of $100/6 = 16.67$ would be obtained. This is the idealised situation. As will be seen below, there are problems that may complicate this.

There are several problems associated with this method. In general:

- far lower preconcentration factors are obtained than in some other methods.
- The method is time-consuming and labour-intensive.
- Sometimes difficulty can be experienced in separating the layers. This is especially true if they form an emulsion. Sometimes emulsions can be broken using heat, *e.g.* a hair drier, by centrifugation or by sonication. Any of these methods would add to the time required for preparation.
- An added complication is that in natural waters, some analytes may be bound to organic material, *e.g.* humic acids. These may not be extracted as efficiently as simple inorganic ions. In such a case, either acid destruction of the organic species or photolysis may be required for "total" concentrations of a metal to be determined.
- Not all analytes will require the same pH or even the same chelating agent. Therefore, if the determination of a range of analytes is required, more than one extraction may be necessary.

- Contamination can be a problem because of the numerous sample handling stages. Similarly, if a large number of samples are to be prepared, then significant amounts of glassware need to be pre-cleaned thus, adding to the preparation time.

2.10.3 Solid Phase Extractions

These can be undertaken using various types of solid phase extraction medium and in several different ways. Assorted media may be used including commercial chelating resins, *e.g.* Chelex-100, Metpac-CC-1 and Muromac A-1. These all have iminodiacetate functional groups and are adept at retaining doubly charged ions. They are particularly good at retaining transition metals, lanthanides, Cd, Pb, U and Zn. They also have some capacity for alkali and alkaline earth metals. These are not usually on the list of analytes and may cause problems by blocking the active sites of the resins, preventing the analytes from being retained. They may be removed from the column by using an ammonium acetate buffer at pH 5.5. Other elements, *e.g.* Cr and Sn, are also retained on the column, but are not efficiently eluted. Commercially available cation and anion exchange materials have also been used, as have numerous resins made in-house. The solid phase extractions may be performed in many ways including the stirring of loose resin with the buffered sample. The resin may then be filtered off and eluted, and the eluate is then analysed. If an analytical technique capable of analysing solid materials is directly available, *e.g.* XRF, the resin may be analysed directly. Another method is to use solid phase extraction cartridges, where the samples are simply loaded into the top of the cartridges. The analytes are retained on the medium whilst the sample matrix is removed to waste (often under vacuum) and then after a known volume of sample has been introduced, a brief washing stage is often used to ensure all matrix and concomitants are removed. The analytes are then eluted using a small volume of acid. If performed under vacuum, the process can be very quick, with the analytes from 10 mL of sample being loaded onto the medium within a couple of minutes. Elution with 0.5 mL of acid would provide a 20-fold preconcentration. The advantage of this is its rapidity and the reasonable preconcentration factors obtained. There are several disadvantages. One is the cost of the cartridges, although some can be re-used, although the analyst should do a test for longevity rather than just hoping for the best. Since some manifolds contain the facility to hold 10–12 cartridges, the analyst should also be very careful to maintain traceability, *i.e.* ensure that the samples are eluted into the correctly labelled vials. Other disadvantages include the possibilities of incomplete retention of the analytes on the medium and, once retained, inefficient elution. Careful optimisation of sample pH and the strength, type and volume of eluent is therefore required. In common with the liquid–liquid extractions, if the analyte(s) are present in more than one chemical form, some species may be retained whereas others, may not.

An alternative to solid phase extraction cartridges is resins placed in a column. These may be purchased commercially (*e.g.* Metapac CC-1 and Muromac A-1) or prepared in-house by placing loose resin in glass or plastic columns with glass wool plugs to prevent the resin from escaping. These may be coupled directly with the detection system allowing preconcentration and matrix removal simultaneously. The preconcentration factor depends on the time the sample is flowed through the resin and the size of the eluent loop. However, potentially high preconcentration factors are obtainable in a relatively short time period. This greatly diminishes the possibility of mis-labelling samples and has the added advantage of being an enclosed system, *i.e.* it also diminishes the possibility of contamination. Of the chelating resins mentioned previously, Chelex-100 is available in numerous particle sizes. However, if the particle size is too small, *e.g.* 400 mesh, a high back pressure results, making it hard for simple flow injection equipment to function properly. Chelex-100 is also the least cross-linked of the materials, meaning that as the sample loading and eluting cycles progress, the resin swells and then shrinks again. This can lead to channelling through the resin bed and the formation of air gaps. These will lead to broadening of analyte detection peaks, hence making measurement more difficult. The commercially available columns Metpac CC-1 and Muromac A-1 are more highly cross-linked and so do not suffer these problems. Numerous other chelating resins have been made in-house.

Several reviews of assorted materials have been presented. These include one that focused on the use of functionalised SBA-15,[13] graphene and graphene-based materials,[14] porous monolithic materials[15] and advanced functional materials (including assorted nano-materials, porous materials such as monoliths and metal organic frameworks, ion imprinted polymers and assorted magnetic materials.[16] These on-line column systems have similar disadvantages as the cartridges. Again, longevity tests should be undertaken on the resins to ensure analyte retention/elution efficiency does not decrease with use. The on-line versions lend themselves more easily to automation, thus freeing analysts to perform other tasks and are therefore often the preferred methodology to adopt.

2.10.4   Other Preconcentration/Matrix Removal Methods

A large number of liquid micro-extraction methods have been developed recently. These are often variations on a theme, but are usually very efficient, require low sample volumes and provide high preconcentration factors. Since there are so many modes of liquid phase microextraction, a detailed description of each will not be given. Instead, review papers that provide the theory, applications, advantages and future outlooks will be given. An example, by Al-Saidi and Emara, discussed dispersive liquid–liquid microextraction (DLLME).[17] Although from 2014, the review gave a decent perspective, naming many of the chelating agents used and giving the preconcentration factors for each of the applications discussed (which were between 42 and 388, depending on the analyte and extraction system). The limitations of each of the systems were also discussed. An extensive review (with 147 references), by Hu *et al.* from 2013, gave a review of liquid phase microextraction for the analysis of trace elements and their speciation.[18] The different modes discussed were: single drop microextraction (SDME), hollow fibre–liquid phase microextraction (HF–LPME), dispersive liquid–liquid microextraction (DLLME) and solidified floating organic drop microextraction (SFODME). A comparison of the performance of each of the modes was given followed by sections discussing the application to environmental samples, biological samples and other samples. In many of the modes, the sample pH, stirring time, ionic strength and flow rate are all critical parameters for obtaining optimal results. The organic solvents should also have low water solubility.

A review, containing 71 references, of other reviews was presented by Rutkowska *et al.*[19] in 2019. The review is split into useful sections discussing historical aspects and basic principles, reviews focused on dispersive liquid–liquid microextraction, reviews focused on specific samples, reviews focused on the specific analytes, reviews focused on specific detection techniques and reviews focused on automation of LPME. It would be an excellent place for somebody new to the area to start, with the references within it giving an enormous amount of information on methods that have been developed. Another useful review was presented by Aguirre *et al.*[20]

For HF–LPME, the hollow fibre used is usually polypropylene (but should certainly be hydrophobic) and often has an inner diameter of approximately 600 µm, which is compatible with the µL volumes of the acceptor solution required for microextraction. The sample should be stirred to ensure that fresh sample passes over the fibre continually. Since the acceptor solution is within the fibre, high stirring speeds are possible, thereby shortening the extraction process. However, it should be noted that too vigorous stirring can lead to air bubbles adhering to the fibre. These would decrease the extraction efficiency. Both SDME and HF–LPME require lengthy extraction times. However, DLLME requires no specialised equipment, with only a microsyringe, a centrifuge and some conical centrifuge tubes being required. The extraction is extremely rapid, simple and uses very low volumes of sample. In addition, it has a low cost and usually provides a high recovery and preconcentration factor. However, DLLME has some disadvantages. These can include poor reproducibility, and it also has a relatively poor ability to remove interferences. Additionally, some problems in the selection of extraction solvent and disperser solvent can occur. Transferring the obtained sedimented phase for introduction to the detection instrument may also prove problematic. The SFODME method is extremely dependent on temperature where it relies on a solvent that is in liquid phase at slightly elevated temperature, *e.g.* 60 °C where it can be mixed with the sample, but then solidifies on cooling to room temperature. It should be noted though that a temperature of <60 °C is often required because of increased solubility of the solvent in the aqueous sample.

Cloud point extraction (CPE) has also gained significant interest in recent years. A good recent review of the subject was presented by Mandal and Lahiri.[21] Briefly, the analytes can be preconcentrated or extracted using varying modes of CPE through three simple steps. These are solubilization of the analytes, cloud formation and then phase separation. The distribution or partition of analyte between the surfactant rich phase and the water rich phase starts above a critical temperature known as the cloud point temperature. The surfactant rich phase plays the role of the organic phase in conventional liquid–liquid extraction. The method offers good preconcentration factors, is generally regarded as being environmentally friendly and may offer the possibility of speciation (*i.e.* it suffers the drawback of not extracting all chemical species of some elements!). A further limitation is that after the process occurs, the analytes end up in an organic solvent. This is fine if detection is to be through the use of FAAS (although the high aspiration rate will limit either the number of analytes to be determined or the preconcentration factor) or ETAAS. The introduction of organic solvents to ICP-based instruments can be problematic since they can de-stabilise the plasma. This may be overcome to a large extent in ICP–OES by using a chilled spray chamber. If this is set at $-5\ ^{\circ}$C, the vapour pressure of the solvent decreases, hence less reaches the plasma. Several specialised "organic" sample introduction systems are available commercially. Introduction to an ICP–MS instrument can be more problematic though. As the solvent desolvates, it decomposes forming carbon particles. These can clog the cone orifices and coat the ion lens system. Even if a specialised organic sample introduction system is used, extended use can sometimes lead to excessive signal drift. A potential way of circumventing the problems associated with the introduction of organic solvents is to evaporate them away and then taking up the residue in an acid. Although this could be introduced far more readily into the instrument, it would clearly increase the sample preparation time significantly and could potentially lead to the loss of volatile elements such as Hg.

With many of these techniques, the volume of extracting liquid is very small to maximise the preconcentration factor. Since a standard ICP nebuliser draws up a sample at $\sim$1 mL min^{-1} and a flame AAS at $\sim$8 mL min^{-1}, these instruments are not usually used directly. Instead, a method of discrete volume injection is used. This may be through a flow injection manifold or through the use of a microsyringe injecting the sample into ETAAS or ETV–ICP–MS or ETV–ICP–OES. A recent review of flow injection liquid phase microextraction containing 98 references was presented by Lemos *et al.*[22] All of the modes of liquid microextraction described previously were also discussed in this review.

2.11 Concluding Remarks

For a successful analysis to occur, any sample preparation procedure requires that the analyte is determined in a manner that is representative of the way it is in the sample. Therefore, contamination must always be minimised. This can start during sampling. For instance, if the roots of a plant are to be analysed, it is necessary to remove all traces of the soil. During the preparation scrupulous acid washing of all glassware prior to use is required. If the analyte is to be extracted from the sample, it should ideally be done so with 100% efficiency. For some sample types, *e.g.* many environmental ones, this will limit the sample preparation methods available. If, however, a "pseudo-total" is required, it may be possible to use a more readily available, safer and easier method, *e.g.* acid extractible totals. For liquid samples, extraction efficiency should again ideally be 100%.

If not, the efficiency should at least be consistent, enabling a correction factor to be applied. A method will not be fit for purpose if three replicate extractions yield efficiencies of 85, 67 and 93% giving an overall value of 81.7%. The precision is too poor to state that the method is under scientific control. However, if the replicate extraction efficiencies are 93, 91 and 94%, an overall correction factor of 1.08 may be applied to the average concentration. This correction is often frowned on analytically though. Instead, it is often safer to state the original concentrations determined and that extraction efficiency was only 92%. If a method requires numerous sample handling steps, the traceability must be kept at all times to ensure that sample confusion does not occur. If a process can be automated then it may, initially, be a nuisance to develop, but would offer significant time and hence cost savings in the longer term.

References

1. Z. Hu and L. Qi, Sample Digestion Methods, in *Treatise on Geochemistry*, ed. K. Turekian and H. Holland, 2nd edn, 2014, vol. 15, pp. 87–109.
2. https://cem.com/en/microwave-digestion#sample_types.
3. S. N. H. Bokhari and T. C. Meisel, *Geostand. Geoanal. Res.*, 2016, **41**, 181–195.
4. P. Goodall, M. E. Foulkes and L. Ebdon, *Spectrochim. Acta, Part B*, 1993, **48**, 1563–1577.
5. L. Ebdon, M. Foulkes and K. Sutton, *J. Anal. At. Spectrom.*, 1997, **12**, 213–229.
6. U. M. F. M. Cerqueira, M. A. Bezerra, V. A. Lemos, J. J. Coutinho, C. G. Novaes, S. A. Araujo and L. P. Do Nascimento, *TrAC, Trends Anal. Chem.*, 2023, **167**, 117277.
7. J. L. Martins Viana, A. A. Menegario and A. H. Fostier, *Talanta*, 2021, **226**, 122119.
8. R. Clough, C. F. Harrington, S. J. Hill, Y. Madrid and J. F. Tyson, *J. Anal. At. Spectrom.*, 2021, **36**, 1326–1373.
9. S. L. C. Ferreira, J. B. de Andrade, M. D. A. Korn, M. D. Pereira, V. A. Lemos, W. N. L. dos Santos, F. D. Rodrigues, A. S. Souza, H. S. Ferreira and E. G. P. da Silva, *J. Hazard. Mater.*, 2016, **145**, 358–367.
10. M. D. A. Korn, J. B. de Andrade, D. S. de Jesus, V. A. Lemos, M. L. S. F. Bandeira, W. N. L. dos Santos, M. A. Bezerra, F. A. C. Amorim, A. S. Souza and S. L. C. Ferreira, *Talanta*, 2006, **69**, 16–24.
11. X. W. Zhao, N. Z. Song, W. H. Zhou and Q. Jia, *Cent. Eur. J. Chem.*, 2012, **10**, 927–937.
12. A. Fisher and D. Kara, *Anal. Chim. Acta*, 2016, **935**, 1–29.
13. A. Larki, S. J. Saghanezhad and M. Ghomi, *Microchem. J.*, 2021, **169**, 106601.
14. W. Q. Jing, J. Q. Wang, B. Kuipers, W. T. Bi and D. D. Y. Chen, *TrAC, Trends Anal. Chem.*, 2021, **137**, 116212.
15. J. C. Masini, F. H. do Nascimento and R. Vitek, *Trends Environ. Anal. Chem.*, 2021, **29**, e0012.
16. M. He, L. Huang, B. S. Zhao, B. B. Chen and B. Hu, *Anal. Chim. Acta*, 2017, **973**, 1–24.
17. H. M. Al-Saidi and A. A. A. Emara, *J. Saudi Chem. Soc.*, 2014, **18**, 745–761.
18. B. Hu, M. He, B. Chen and L. Xia, *Spectrochim. Acta, Part B*, 2013, **86**, 14–30.

19. M. Rutkowska, J. Płotka-Wasylka, M. Sajid and V. Andruch, *Microchem. J.*, 2019, **149**, 103989.
20. M. A. Aguirre, P. Baile, L. Vidal and A. Canals, *Trends Anal. Chem.*, 2019, **112**, 241–247.
21. S. Mandal and S. Lahiri, *Microchem. J.*, 2022, **175**, 107150.
22. V. A. Lemos, R. V. Oliveira, W. N. L. dos Santos, R. M. Menezes, L. B. Santos and S. L. C. Ferreira, *TrAC, Trends Anal. Chem.*, 2019, **110**, 357–366.

3 X-ray Fluorescence Spectrometry

3.1 Introduction

X-ray fluorescence (XRF) spectrometry is one of the most established instrumental elemental analytical techniques. It was developed in the 1920s, but it was only in the 1940s that detectors became sufficiently well-developed for it to become a useful technique. As with other chapters in this book, the theoretical background (in this case, underlying the XRF effect) will not be given in any great detail, because that will be covered in another publication of this series. However, a brief summary of the fundamentals as well as some of the types of potential interferences is necessary to aid the reader.

As with many other methods of atomic spectrometry, the fluorescence effect involves the transition of electrons from one energy state to another. However, this occurs much deeper in the atom, *i.e.* closer to the nucleus than other methods. The process is depicted in Figure 3.1. As an incident X-ray enters an analyte atom, it has sufficient energy to both excite and expel an electron from one of the inner orbitals, often identified in terms of their shells, *e.g.* K, L, M, *etc.* This leaves a "hole", effectively a vacancy in the high energy inner shell electron structure, making the atom unstable. To remove the instability, this "hole" is back-filled by one of the electrons from a higher energy orbital, *e.g.* from an L shell to a K shell, and, as this transition occurs, energy is released in the form of an X-ray. Each of

Practical and Technical Guides for Laboratory-based Chemists No. 1
Atomic Spectrometric Methods of Analysis
By Andrew Fisher
© Andrew Fisher 2025
Published by the Royal Society of Chemistry, www.rsc.org

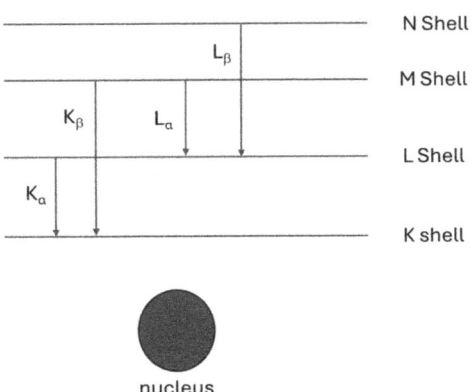

nucleus

Figure 3.1   A schematic describing the nomenclature of XRF lines.

the transitions will be characteristic of the individual analyte. The lines are usually described by a name rather than an energy value. Therefore, if the "hole" appears in the K shell, it will be called a K transition. If the back-filling electron originates in the L shell, the transition will be called K_α (or K–L). If it originates from an M shell, it will be termed K_β (or K–M). Similarly, if the "hole" is in an L shell and the back-filling electron comes from an M shell, it will be termed L_α (or L–M), from an N shell L_β (or L–N), *etc.*

3.2   Advantages and Disadvantages of Energy Dispersive and Wavelength Dispersive XRF

There are two instrumental categories of XRF spectrometry: wavelength dispersive (WDXRF) and energy dispersive (EDXRF). The relative advantages and disadvantages of each are summarised in Table 3.1.

The advances in XRF instrumentation, methods of interference correction and new applications are reviewed on an annual basis in the Atomic Spectrometry Updates series published in the Royal Society of Chemistry journal *"Journal of Analytical Atomic Spectrometry"*.[1] While the one referenced here specialises in XRF techniques, other updates may also have interesting applications. Other reviews on specific areas also exist. An example, by Margui *et al.* reviews the use of XRF spectrometry for environmental analysis.[2] This review provided the reader with the basic principles, the instrumentation required, and some applications and gave a perspective on possible future trends.

Table 3.1 Advantages and disadvantages of EDXRF and WDXRF spectrometry.

	EDXRF	WDXRF
Elemental range	F–U	Be–U
Detection limit	Low $(mg\,kg^{-1})$ for heavy elements. Less good for lighter elements (tens of $mg\,kg^{-1}$)	Low $(mg\,kg^{-1})$ for most elements
Sensitivity	Good for heavy elements. Less good for light elements	Reasonable for light elements. Better for heavy elements
Resolution	Good for heavy elements. Less good for light elements	Good for light elements. Less good for heavy elements
Cost	Relatively inexpensive	Much more expensive
Power consumption	5–1000 W	200–4000 W
Measurement	Simultaneous	Sequential/simultaneous
Moving parts	None	Crystal goniometer
Speed of analysis	Seconds – a few minutes	Depends on the mode of analysis, but can be up to nearly an hour

3.2.1 Interferences

There are several modes of interference that can cause problems during XRF analysis. Broadly, they may be divided into non-spectroscopic and spectroscopic effects. The "critical depth" is the region that the X-rays analyse. This region may only be a few microns deep, and for bulk samples, it is dependent on the density, mass attenuation coefficients (of elements involved) and characteristic energy of the X-ray. It is defined as the depth below the irradiated surface beyond which 99% of an element's X-ray line emission is undetectable due to being absorbed in the sample. In Figure 3.2(a), there are an equal number of black and grey atoms. However, if the sample is not homogeneous, as in Figure 3.2(a), then the number of grey atoms in the "measuring zone" is an over-estimate of the total, whereas the number of black ones is an under-estimate. These effects can be minimised, especially for environmental samples, by milling the material to a very fine powder and then mixing further. Some sample types may be homogeneous but others, *e.g.* alloys, may have inclusions that may be problematic. Figure 3.2(b) shows the effects of an uneven surface. Here, problems of "shadowing" may arise, *i.e.* an atom (or particle) can be masked by another. Again, over- and under-estimates can result. The extent of these problems depends on the sample type and on the preparation method. The latter will be discussed at greater length in the following sections. Figure 3.2(c) shows

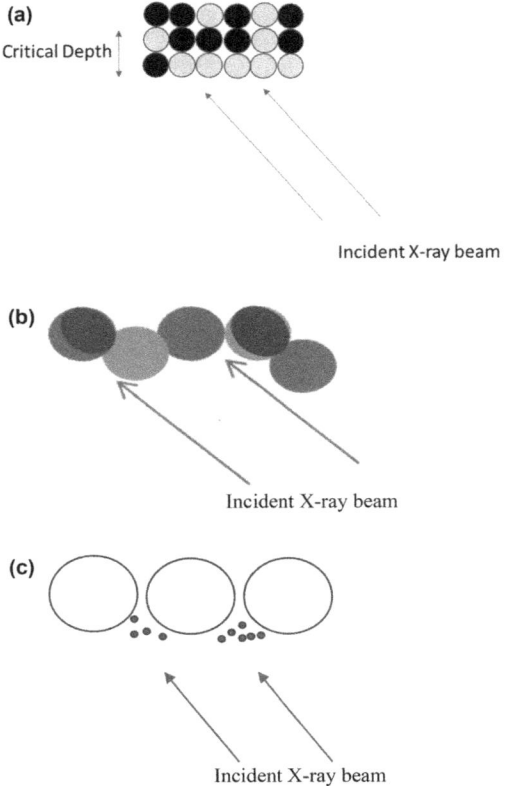

Figure 3.2 Types of interference in X-ray fluorescence analysis.

particle size effects. An inhomogeneous sample containing small particles and large particles could see the smaller ones sinking to the bottom of the sample cell resulting in an over-estimate of their number. The extent to which this happens is dependent on the sample preparation method but can be appreciable with loose powders. All of the above examples arise through errors or problems with sample preparation and may be overcome or minimised using some care.

There is a different type of interference and this is demonstrated in Figure 3.3. This shows the principle of inter-element effects and is very sample specific. It is a problem that is not so easily overcome and requires 'selective' interference correction techniques to obtain reliable data. A higher energy element, *e.g.* Fe, fluoresces X-rays which then excites a lower energy analyte, *e.g.* Cr. The Cr signal is therefore enhanced and the Fe signal diminished. The particular problem shown in Figure 3.3 is potentially the greatest during the analysis of

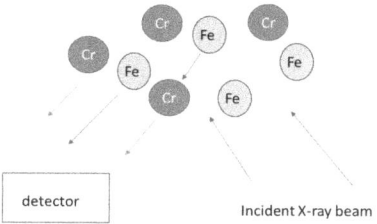

Figure 3.3 Inter-element interferences.

steel samples, but other examples exist for other analytes/sample types. Interference correction could include fundamental parameters and empirical methods. How these methods work is outside the scope of this text, but analysts need to be aware of their existence.

Many manufacturers supply their instruments with a selection of pre-programmed analysis protocols. This is potentially tailored to the applications undertaken in the laboratory to which the instrument is supplied. However, it is also possible for analysts within any laboratory to programme the analysis procedure required. When doing this, it is important to have some knowledge of the physical nature of the samples under analysis so that potential interferences can be avoided (or minimised). Numerous examples exist and a selection of them are itemised in Table 3.2. The potential interferent may come from concomitant elements in the sample but may also originate from the X-ray tube. For instance, the Fe K_α line at 18.307 Å can be interfered with by a Rh L_α line (at 18.377 Å) originating from a Rh-based tube. It is important that an analyst understands the likelihood of interferences for the analytes required so that they may choose an

Table 3.2 Selected potential interferences during XRF analysis

Analyte line and wavelength (Å)	Interference line and wavelength (Å)
Ca K_α at 3.360	Sn L_β at 3.385
Ti K_α at 2.750	Ba L_α at 2.776
Fe K_α at 1.937	Mn K_β at 1.910
Co K_α at 1.791	Fe K_β at 1.757
Rb K_α at 0.927	Bi L_β at 0.926
Zr K_α at 0.788	Sr K_β at 0.783
Mo K_α at 0.710	Zr K_β at 0.701
Sn K_α at 0.492	Ag K_β at 0.497
Sb K_α at 0.472	Cd K_β at 0.475 and Sn K_α at 0.492
Pb L_α at 1.175	As K_α at 1.177

appropriate wavelength. For instance, if an analyte, *e.g.* Co is to be determined in a soil sample, then it is likely that Fe will be present as a concomitant at significant concentrations. It would therefore be unwise to use the Co K_α line at 1.791 Å because of the potential interference from the Fe K_β line at 1.757 Å. Although this is by no means the most sensitive Fe wavelength, it is more than sufficient to exert a potential interference effect, hence leading to an overestimate of the Co content.

3.3   Methods of Sample Analysis

Both EDXRF and WDXRF techniques are capable of analysing solids, powders and liquids, although the analysis of liquids is performed by only a small minority of laboratories. Liquid and powder samples should be placed in specialised cups and usually have specific places on the autosampler tray. There are usually far fewer places on the autosampler for specialised sample cups and therefore the analyst is less free to undertake other tasks because they must return periodically to change the samples.

3.3.1   Liquid Samples

As discussed previously, there are relatively few laboratories that analyse liquid samples. That having been said, there is no reason why materials such as oils, water or even concentrated acid solutions should not be analysed. Most liquid sample types require no preparation other than to be placed in a specialised cell. The cell has a film bottom, through which the X-rays pass. A variety of films exist, depending on the sample type. They could be Mylar (a polyester) or polycarbonate (which are suitable for the analysis of organic liquids, *e.g.* gasoline, oils and some solvents) or polypropylene (suitable for acids). The thickness of these films can vary, depending on the material, but can be up to 12 μm. It should be noted that the thicker the film, the more attenuation of analyte X-rays – especially for the lighter elements. Different types of cells are also available. Some just require a lid to be placed on top whereas others require assembly. The cups are designed to be safe, *i.e.* to prevent leaks of the sample into the XRF instrument. A liquid landing on the X-ray tube could destroy it. Such analyses should therefore only be undertaken by trained users or under their supervision while a novice learns the SOP. Care should be taken to ensure that the material from which the cup is made is compatible with the sample. It would be catastrophic if the cell starts to dissolve or deform in the presence of the liquid sample. To ensure that disaster will not happen, a suitable time lag between preparation and analysis should occur. This will identify if the cell is going to leak. The cells should never be re-used. In common with the analysis of loose powders, the analysis cannot be performed under vacuum. Instead, a helium atmosphere should be used, although some EDXRF instruments may allow the analysis to be performed in an air or nitrogen atmosphere.

There are four common methods of analysing solid materials. These are: loose powders, pressed pellets, fused glass disks and, for environmental samples, particulates trapped on a filter. Each of these will be discussed below, giving practical hints and tips to aid their analyses.

3.3.2 Loose Powders

Here, a sieved and milled sample is placed in a demountable container that has a thin X-ray-transparent polymer film stretched across the base. The polymer is usually (but not always) a polyester and goes under several tradenames, *e.g.* Mylar, Melinex and Hostaphan. These films are available as pre-cut circles that are already cut to size or as a roll that the analyst can cut to size themselves. An example of a loose powder cell is shown in Figure 3.4.

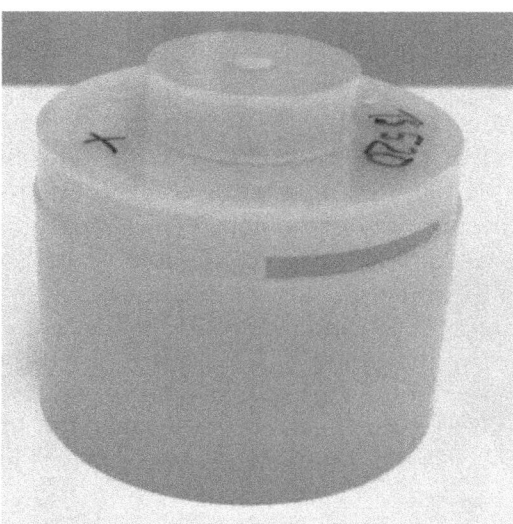

Figure 3.4   An example of a loose powder cell.

This is the quickest and easiest of the sample preparation methods, with the construction of the sample holder containing the sample taking only a couple of minutes. To minimise problems arising from different sized particles, the sample should still be dried and milled. If it is not milled, analysis is still possible, but potential errors will be maximised.

3.3.2.1 Advantages and Disadvantages

As mentioned previously, it is the quickest method of sample preparation/presentation. It also has the added bonus that the sample may be readily recovered for further testing if necessary. The sample does not have to be weighed accurately into the cell. As long as there is sufficient to cover the polymer sheet at the base of the cell, then that should be sufficient. The loose powder method does, however, suffer from the most interferences. These may be particle size oriented, *e.g.* smaller particles sinking to the bottom of the cell, hence potentially being over-represented during the analysis. Alternatively, if larger particles have reached the bottom of the cell, they may cause "overshadow", *i.e.* they absorb the X-rays and prevent them from reaching the smaller particles. Although these two phenomena have opposite effects, an analyst can clearly not assume that they cancel each other out. The magnitude of the problem is sample specific, with some types that are well-defined, *i.e.* an industrial sample where the matrix is homogeneous, will be a lot less of a problem than an environmental sample where an assortment of different minerals and particle sizes may be present. Another disadvantage of the loose powders is that if the polymer film ruptures, the sample is lost into the instrument, where it may coat the X-ray source or the detector. Cleaning it off is not an easy task and risks damaging them and, since they are extremely ex-pensive components, any error could be catastrophic financially. Loose powders are analysed under a flow of helium rather than under vacuum like most analyses. This adds to the cost of the analysis. It is unwise to analyse the same sample twice without re-constructing the demountable cell with a fresh polymer film in between. Multiple analyses of the same sample are not common for most samples, but if a certified reference material is used, there is a greater possibility. The containers are not air-tight and there could potentially be an ingress of water vapour over time. Therefore, the cells should not be prepared and then left for a substantial time before analysis.

Figure 3.5   A pressed pellet.

3.3.3   Preparation of Pressed Pellets

This is a very common method of sample preparation, especially for geological or environmental materials. It is relatively quick (compared with fused glass disks) and does not suffer the drawbacks of analyte loss through volatilisation. A pressed pellet comprises the particulate sample homogeneously dispersed in an organic binding agent and compressed into shape. Numerous binders have been used including stearic acid, mannitol, and assorted waxy substances. Several trademarked products are now available that offer good binding ability while also being sufficiently pure. Pressed pellets are often up to 40 mm in diameter. An example is shown in Figure 3.5.

A possible method for the preparation of a pressed pellet is outlined below. This may be adapted for other sample types and mill designs.

- Add a sieved (<180 μm) sample (20 g) directly to each grinding bowl of the mill.
- For traceability, note the grinding bowl code against the sample ID because otherwise, once the mill spins, it will be impossible to identify which sample is which.
- Place eight 10 mm agate balls in each 80 mL bowl (for 12 mL bowls place 4 g sample in each with four 10 mm balls)
- Mill for 3 minutes at 300 rpm
- After 3 min, empty the sample onto a sheet of clean, white paper or plastic in a dust cabinet and brush out the bowl with a plastic bristle brush and vinyl-coated spatula if necessary.

- Weigh 12 g of the milled sample into a container. Keep the remaining milled sample for a fusion and/or LOI test if necessary (labelled plastic bag).
- Weigh 3 g of binder (*e.g.* Ceridust 60 50 M, or similar) into the same container until the total mass is 15 g $\pm$ 0.01 g
- Place the 15 g back into the grinding bowl and mill for a further 3 minutes at 300 rpm
- Empty the contents into a labelled container and clean the bowl and beads ready for the next sample (the method for this is given below).
- Weigh 8 g of the mixture of sample and binder into a 40 mm diameter aluminium dish (these are available commercially for XRF pellet formation). Do this in stages ensuring the sample is gently tamped down so that no air gaps remain. When it has all been added, a small mound of material should protrude above the rim in the centre of the dish.
- Place the dish on the bottom portion of the die's base-plate and assemble the rest of the pressing die around it. Ensure no sample is spilled.
- Place the pressing pin (piston) into the piston chamber and let it fall into place.
- Press at ~150 kN for about 30 seconds.
- Release the pressure, dismantle the die and label the underside of the aluminium dish with the sample name using an indelible marker pen.

Cleaning of the grinding elements is necessary because the binder is "sticky". Addition of sand to the bowl and grinding balls followed by a period of 10 minutes of grinding at 300 rpm should remove most of the sample/binding agent from the surfaces. This can then be discarded. A repeat grinding with sand may be necessary for very stubborn samples. Once "clean", the bowls and grinding balls may be washed and dried ready for the next use.

Grinding balls and the inner surface of the bowls are usually made of natural agate or a synthetic equivalent. This is a relatively hard material (about 6.5–7.0 on the 'measurement of hardness' also known as the Mohs scale; where 1 = talc and 10 = diamond). However, since the scale goes up to 10, there are harder materials. This means that if a sample that contains some of these very hard materials is ground, then the sample is more likely to grind (and hence damage) the grinding balls and bowls and contaminate the sample. Although other grinding ball materials are available to use that have greater hardness, it should be stressed that agate is adequate for the vast majority of applications, including environmental/geological ones. It should be noted that if bowls are made of natural agate, then natural agate grinding balls should be used. Similarly, synthetic agate bowls should be combined with synthetic agate balls. Inter-mixing is unwise. As an aside, it should be noted that the Mohs scale is not an acronym of "measure of hardness scale". Instead, it is a scale devised by Friedrich Mohs.

As can be seen, the process is very time-consuming and requires significant amounts of samples. There are ways of making it faster and using less sample though. These are itemised below:

A more rapid way of mixing the milled sample and binder is to weigh the correct amounts into a plastic centrifuge tube, adding two or three of the grinding balls and securing the lid. The tube may then be placed on a laboratory shaker for 5–10 minutes, or until no more white flecks of un-mixed binder can be seen. Although still not rapid, it does save the extra stage of using sand to remove the remnants of sample and binder from the grinding bowls. It is, however, a much gentler approach to mixing and so there is a risk that particle size reduction and mixing may not be complete.

If there is a limited sample amount, *e.g.* only 4 g, then it can be milled and mixed with 1 g of binding agent in the normal way. However, a layer of sand may be placed in the bottom of the aluminium dish to help pack the sample out. The mixture of sample and binder may then be added to the top of the sand. Since the X-rays do not penetrate more than a few microns (the critical depth), the layer of sand will not be analysed – assuming that the sample, binder and sand are not all mixed together.

3.3.3.1 Potential Errors

When assembling the pressing die, it is imperative that the piston be allowed to drop under its own weight to the sample surface. If it is forced down, it may be going in at a slight angle which will lead to a poorly pressed pellet, wasted sample and potentially, a damaged face to the piston. The pellet face should be as flat as possible and so if the piston end becomes damaged, this damage will be imprinted on the pellet face, potentially causing analytical error. If the face of the piston is damaged, it may be sent away to be cut and polished again. This is expensive (although less so than buying a new die set) and has the added inconvenience of preventing the analyst from using it while it is away.

It is also imperative that the binder and sample powders be mixed thoroughly. Obviously, if large flecks of binder remain, then a non-representative sample will be presented to the XRF instrument.

Once the pressed pellet has been prepared, there is often a small amount of powder attached to the sides. This has the potential to become dislodged during analysis. If that does occur, it will fall into the XRF spectrometer and perhaps start coating vital surfaces, *e.g.* the X-ray tube or the detector. Over time, this will lead to significant sensitivity drift. It is therefore safer to scrape excess sample off the sides of the pellets before analysis. Care should be taken not to damage the pellet surface.

The sample surface of the pellet should not be touched with fingers as they will leave a deposit, *i.e.* a fingerprint. Since X-ray analysis is a quasi-surface analysis technique, the fingerprints will affect the results significantly.

Although technically not really an error, it is possible that the pellet can fall out of the aluminium dish. This happens occasionally if the sample has extremely low levels of natural cohesive material, *e.g.* organic matter. It should be noted that if this happens during a lengthy run of samples, then the pellet will remain in the instrument's analysis chamber with subsequent pellets being placed on top of it. The same sample will therefore be analysed numerous times rather than the other samples, and hence, the analysis will have to be re-started from the last valid measurement. Depending on how badly the pellet has been damaged, it may be fine to accept the data as produced or, it may require the sample to be prepared again. This may not require the use of a further 8 g of material. Instead, the existing pellet may be broken up and mixed with an extra 0.5 g of binder and then pressed again.

3.3.3.2 Advantages and Disadvantages

The pressed pellet method of sample preparation has several advantages and disadvantages. An advantage is that it does not dilute the sample overly (8 g + 2 g of binder). Therefore, the limits of detection for the raw material are better than the fused glass disk method. If insufficient sample is acquired, the remedial action described above may be employed. This is not the case with the fused glass disks. It is a cheaper methodology than fused glass disks because it does not require expensive platinum ware or specialised fusion furnaces.

The main disadvantage is that of interferences. While the pressed sample face looks perfectly flat, at the microscopic scale, it is not.

Therefore, the problems of surface roughness, larger particles "shadowing" smaller ones, *etc.* exist to a greater extent than experienced using the glass disk method. However, they are much improved over the 'loose powders' method. Another drawback is that the sample is irreversibly mixed with a binder and hence cannot be recovered for further experimentation. Preparing a pressed pellet requires more sample than many of the other methods. Although the method can be modified to reduce the amount required, it is still likely to be more than that required for fused glass disks or loose powders.

Once prepared, pellets remain fairly stable as the binder tends to be water-repellent. However, the longer they are kept, the greater the chance of accidental damage to the sample surface unless they are stored in a safe environment, *e.g.* a petri dish.

3.3.4   Preparation of Fused Glass Disks

This method is most frequently used for the analysis of geological samples such as rocks, tephra, soils and sediments, but may be adapted to other materials such as glasses, ceramics, metals and salts. Many instrument manufacturers provide very useful booklets or webinars regarding fusions. These are useful places to start for a novice analyst. In some ways, it is the best way to prepare a sample because the original sample structure is completely annihilated. This means that it overcomes many of the matrix effects, *e.g.* sample particle size effects, shadowing of larger particles over others and provides a smooth, homogeneous surface for the analysis. An example of a glass disk is shown in Figure 3.6.

Figure 3.6   Example of a fused glass disk.

The method uses expensive platinum ware (typically 95% platinum and 5% gold) and so care of this is an important factor. In general, a great deal is known about most sample types, *e.g.* if XRF instrumentation is to be used as a quality control measure, the analyst already has a reasonable knowledge of the components of the samples. However, environmental samples may be more problematic, since knowledge of the local geology from where they came may be unknown or incomplete. This may lead to catastrophic damage (see Figure 3.7) if precautions are not taken. This is because if they contain high concentrations of platinum poisons or materials that combine with the platinum, once the temperature is raised, the platinum may boil away. This effect can be facilitated by conditions which promote reduction of the sample components to their lower oxidation or even their elemental state when processed at elevated temperatures. There are numerous examples. For instance, high concentrations of Co, Cr, Cu, Fe, Mn, Ni and Zn can combine with platinum alloys at temperatures above 1200 °C. At relatively low concentrations, they can leach into the platinum ware and then leach out again into subsequent samples, *i.e.* they can cause cross-sample contamination. At higher concentrations, they can lead to cracks and, in extreme cases, holes in the platinum. Maintaining temperature at 1200 °C or below helps minimise these problems. Other elements, *e.g.* Ag, Bi, Pb, Sb and Sn, readily alloy with the platinum, even at relatively low temperatures. Non-metals and some metalloids such as As, P, S and Si may also cause terrible damage. It should be noted though that it is Si as the element that causes the

Figure 3.7 Damaged platinum crucible.

trouble rather than silicates and so the analysis of environmental samples is still possible.

Many of these potential problems may be overcome by maintaining oxidising conditions, *i.e.* choosing the correct flux. Numerous fluxes are available including mixtures with assorted concentrations of lithium metaborate and lithium tetraborate, providing different pH conditions and the ability to fuse different substances. One example is 66:34 lithium metaborate:lithium tetraborate. This is of particular use for the fusion of alumino-silicates (many rocks, soils, sands and sediments) as well as iron ores and nickel. Other specialised fluxes may contain between 5 and 20% sodium nitrate as an oxidising reagent to ensure that all Si remains as the silicate. Many manufacturers will supply a "cookbook" where the ideal flux, temperatures required, ramp rates to reach the temperatures, *etc.* are given for many sample types. The manufacturers of XRF instruments have an applications department employing experienced analysts. This is often a good place to start when developing new methodologies.

Prior to the preparation of fused glass disks, the sample should be thoroughly milled so that it is as fine a powder as possible. If the particles are too large, incomplete fusion will occur, leading to tiny particles of sample being visible in the glass. Once milled, an aliquot of the sample is usually taken for a loss on ignition (LOI) test. This is especially important during the analysis of environmental samples. Such tests provide an insight into the reactive composition of a sample at elevated temperatures and any potential reducing conditions which may need to be overcome. A methodology for determining loss on ignition is given below:

- Precondition the ceramic crucibles at 105 ($\pm$4) °C overnight. New ceramic crucibles may require conditioning at the furnace temperature (*e.g.* 1050 °C) prior to their general use.
- Allow the preconditioned crucible to cool to room temperature in a desiccator and then record the weight using an analytical balance.
- Sample material must be dried at 105 °C ($\pm$4) °C for a minimum of 2 hours prior to placing in the furnace.
- Weigh 1 g ($\pm$0.09 g) of the dried (and milled) sample into a crucible and record the weight to 4 decimal places.
- Record the sample code against the position of the crucible on the muffle furnace tray – **NOTE:-** do not try to label the crucibles themselves as any markings will be burned away during heating.

This could potentially lead to the analyst having no idea which sample is which and would require the procedure to be repeated.
- Place in the furnace and heat at 1050 °C for 1 hour (1000 °C for material with a high concentration of volatile compounds, *e.g.* fluorine; chlorine)
- Remove the heated crucible and place on a heat proof plate.
- Allow to cool to room temperature in the desiccator and reweigh.
- Calculate LoI using the following formula:

$$\text{LoI}(\%) = (W_s - W_A)/(W_s - W_c) \times 100$$

where W_s = weight of crucible + sample dried at 105 °C (g), W_A = weight of crucible + ignited sample weight (g), and W_c = weight of pre-conditioned empty crucible (g).

Occasionally, the sample gains weight during the 'loss on ignition' process. This can be indicative of species such as sulfides being oxidised to sulfates and could mean that the samples are potentially harmful to the platinum ware. If significant increases in weight are obtained, it may be worthwhile doing an analysis of the loose powder prior to preparing the glass disk. This would help in identifying potential harmful species. Alternatively, the flux used may be changed to one that contains an oxidant such as sodium nitrate.

A method for the fusion process is outlined below. The temperature and flux used may change according to the sample type, but the method below is adequate for many sample types.

- Lithium metaborate/tetraborate (66 : 34) flux should be oven dried (minimum 2 hours) at 105 °C (±4 °C) prior to use.
- Sample material should be oven dried to constant mass at 105 °C (±4 °C) prior to use. Samples with suspected high concentrations of volatiles (*e.g.* fluorine; chlorine) should be dried at 40 °C.
- Sample material should typically be in powder form with particle size <63 μm. Milling as described for the pressed pellets should ensure this.
- Samples should be weighed to four decimal places using a calibrated analytical balance.
- Weigh the required quantity of flux directly (9 ± 0.01 g) to a clean platinum crucible and record the mass.
- Tare the balance and weigh the required quantity of sample (0.9 ± 0.01 g) directly to the crucible containing the flux and record the sample mass.
- Gently mix with a clean glass rod or plastic spatula.

- Add 0.3 mL of lithium iodide solution (5% w/v) as a non-wetting agent to the crucible (alternatively, an ammonium iodide pellet may be added – the fusion furnace has ports for this and will introduce the pellets automatically at the correct stage during the fusion).
- Repeat the above five steps for sufficient samples to occupy all of the holders in the fusion furnace. Older versions may have space for two samples, whereas more recent versions have space for six samples.
- Place the crucible and the receiving dish in the holders in the fusion furnace (the crucible in the top holder and the dish in the lower).
- Programme the fusion furnace according to the manufacturer's specifications for that sample type and press the start button.
- The furnace will go through its programme and provide a still hot glass at the end of it.
- Using platinum tipped tongs, remove the dish from the holder and let it cool.
- Label the sample whilst in the dish to ensure that the label is placed on the upper side.
- Gently tap the dish so that the now, cooled glass disk falls out.

The above procedure is adequate for a 40 mm diameter glass disk. If a smaller diameter is required, *e.g.* 32 mm, then at least 6 g of the mixture of sample and flux is necessary (6 g of flux and 0.6 g of sample). This is useful for when there is a limited amount of sample.

Once a total analysis has been completed, the percentage loss on ignition plus all of the major components identified by the XRF instrumentation should add up to $100 \pm 1\%$ of the sample – especially when using WDXRF instrumentation. If the data do not total $100 \pm 1\%$, then something has probably gone wrong and a customer (or a line manager) is likely to request a repeat analysis, including preparation of another disk. Sometimes, it is obvious when something has not gone completely according to plan. Some glasses may not be completely clear. If a darker shadow is visible in the glass, then incomplete fusion has probably occurred. This can potentially be prevented or overcome by ensuring that the milling process has broken the sample down into micron sized particles.

3.3.4.1 Potential Errors

If the sample has not been milled to <63 μm, the larger particles may not undergo complete fusion. This will lead to analytical error (as described above) but may also cause the disk to crack.

The non-wetting agent (ammonium iodide pellet or lithium iodide solution) has to be added in the correct dose. If insufficient is added (or if it is omitted), then the melt may not be poured into the receiving dish as efficiently leaving large amounts of sample in the crucible. In addition, the glass disk may stick to the sides of the receiving dish and then crack as it cools. If too much non-wetting agent is used, the disk formed may not fill the surface of the receiving dish, *i.e.* it is poorly shaped.

The platinum receiver dish should be mirror-like, *i.e.* have no scratches on the surface. If scratches do exist, they will become imprinted on the glass disk and potentially lead to errors during analysis. Small scratches may be removed using a specialised felt cleaning stick, a few drops of machine oil and diamond paste (polishing kit). In general, two or three different diamond pastes may be used; starting with those with larger sized particles, *e.g.* 10 µm and then decreasing in size down to 3 µm. This procedure can be undertaken on a special lathe with jigs designed for dishes and crucibles. The problem with the procedure is that it does remove significant amounts of platinum. A new dish may weigh 45 g, but after numerous cleaning cycles, it may decrease in weight by over 20%. Inevitably, repeated cleaning of the dish will eventually lead to it being unusable as the amount of platinum removed increases. The scratch removal process should therefore only be undertaken when really necessary. If the receiving dish has received a very deep scratch, it may not be possible to remove it using the above technique. It may then be necessary to send the dish away for specialist re-polishing. This procedure is not cheap and also removes larger amounts of the platinum.

Both crucibles and receiving dishes can be cleaned between samples by placing them in a beaker of 20% citric acid in a sonic bath set at approximately 60 °C. This will remove droplets of material in the crucibles that were not tipped out and particles from the rim of the glass disk that become stuck to the receiver dish. Either could potentially contaminate the next sample if not removed. After sonication in the citric acid solution, the crucibles and dishes should be rinsed thoroughly using ultra-pure water and then dried. The cleaning process should be undertaken after each sample preparation to avoid cross-contamination. This adds to the time required for preparation considerably, with the cleaning process taking approximately 10 minutes and the drying process even longer. If the crucible is not completely dry, then an accurate weight of the next material will not be obtained because the mass read will decrease as the water evaporates.

A suitable summary of how platinum ware should be looked after is given in ref. 3.

There are several reasons why a glass disk may crack. These include incompletely fused samples and an incorrect amount of non-wetting agent described above. Other causes include internal stress caused by the incorrect temperature being used during preparation, severe scratches on the platinum receiver dish and thermal shock, *i.e.* a very hot disk coming into contact with something much cooler. Disks do not always crack within a few minutes of them being tipped out of the receiver dish. Sometimes, cracking can occur hours later (a process called the delayed stress effect).

If the temperature used for fusion is too low, the viscosity of the melt will be too high. This will lead to poor spreading of the melt when poured into the receiving dish. This too will lead to a poorly shaped disk, possible incomplete fusion of sample particles and po-tential cracking of the disk.

The sample and flux have to be weighed very accurately. This is a repetitive and often boring part of laboratory work. It is possible that an analyst will have a lapse of concentration and tare the balance at the wrong moment (or forget to tare it). This will obviously lead to errors. There are specialised instruments available commercially that can do the weighing automatically. For a laboratory that specialises in XRF analysis, investing in one of these may potentially prevent this type of error while having the additional advantage of freeing up an analyst to do other tasks.

3.3.4.2 Hints and Tips

There are several small points that may be observed that will help an analyst new to the area.

The first big tip is that if there is a standard operating procedure (SOP) for the preparation of the sample types to be analysed, then this should be read and followed. Cutting corners is likely to lead to long term errors.

After drying the flux and samples at 105 °C, remove them from the oven and place them in a desiccator. Do not leave them on the bench to re-absorb water. Similarly, do not take them from the oven and try to weigh an appropriate amount immediately. Hot materials will unsettle the balance and lead to errors.

Once the crucibles and receiving dishes have been cleaned after a sample, they must be allowed to dry completely. Failure to let them dry will also lead to weighing errors of the next sample into the cru-cible as the water evaporates. Drying may be achieved using a soft tissue or by placing them in the oven at 105 °C (again allowing them to cool before use).

The manual weighing of samples and flux is time-consuming and potentially prone to error. Instrumentation has therefore been developed that undertakes this task automatically, hence reducing labour time by 80–90%. The downside is obviously the additional expense but when taking into account that an analyst is made available to do other tasks, it is money well spent.

When the fusion cycle is complete, the glass disks are cooled by a flow of air. This is a normal part of the program. The analyst should not try to take them out using fingers or even gloved fingers. Although the disks are no longer extremely hot, they are still at a sufficiently high temperature to cause pain. When removing the samples from the furnace, the platinum tipped tongs should be used.

When labelling the samples, a small sticky label placed off-centre is best. A large label placed in the centre of the disk may lead to the autosampler experiencing problems when picking up the disk during analysis. The labels should be placed on the disks while they are still in the receiving dish. This ensures that the correct surface (*i.e.* not the surface to be analysed), is labelled.

Sometimes, the disks are not removed from the receiver dish without some persuasion. This can be achieved by turning the dish over so that the glass disk is underneath it and then gently tapping the dish on a hard surface so that the disk is released. Doing this on a paper towel at least gives a small cushion effect as the disk falls out. Since the disk will land on the side containing the label, contamination does not occur.

Always use gloves when handling the disks to ensure that fingerprints are not transferred to the side to be analysed.

When placing the disks in the autosampler positions, ensure that the surface of the autosampler is clean from debris, *i.e.* particles of glass from other samples or from pressed pellets are not present. If these adhere to the surface of the samples, errors may occur. Alternatively, they may stick to the surface of the sample, but then fall off during analysis and enter the XRF instrument itself.

When programming the instrument, it will require the exact weights of flux and sample to be input. Ensure thorough checking and double-checking that the correct values have been input otherwise severe calculation errors may result.

3.3.4.3 Advantages and Disadvantages of Glass Disk Preparation

The main advantage of this methodology is, as discussed above, that interferences caused by the original sample matrix are minimised.

The sample structure should be completely annihilated, and therefore, no surface roughness, no particle size effects and no shadowing of some minerals over others should occur. Of the solid sample preparation methods for XRF spectrometry, this leads to the analysis suffering the fewest problems.

The drawbacks include the necessity of using expensive platinum ware that can be damaged if care is not taken. Another drawback is that it is an extremely lengthy process. Even if milling of the sample occurs on a previous day, if only a two-position fusion furnace is available, then perhaps only ~20 samples may be prepared per day. This is because the cleaning and drying of the platinum ware takes as long as the fusion process itself.

Other drawbacks include the potential loss of volatile elements and the dilution factor involved (9 g of flux + 0.9 g of sample = 11-fold dilution) which may "dilute" some analytes to below the limit of detection. Another disadvantage is that the sample is completely destroyed and can therefore not be used for any further experiments. Once prepared, the glass disks are not totally inert and can slowly become cloudy as moisture levels increase. Therefore, if samples need to be stored, they can be placed in a vacuum desiccator.

3.3.5 Analysis of Particles on a Filter

This specialised method of analysis is undertaken by those laboratories involved in, for example, environmental monitoring. The method is not used only for environmental monitoring but is certainly a major use. It can also be used for general collection from liquids, air, aerosols/exhaust gases as well as pre-concentration filters to improve LODs, *etc.* It suffers the drawback of requiring a reasonable amount of material to be deposited on the filter and for that material to be spread homogeneously across the filter surface. If only a few particles are deposited, then the incident X-ray beam may not representatively sample (or cover?) the filter surface. This could obviously lead to poor precision and accuracy. If environmental particles are to be analysed for a suite of elements including Al and Si, then glass fibre filters should not be used. This type of filter is prone to contamination anyway, and so they should be used with some caution even if Al and Si are not amongst the analytes of interest.

3.4 XRF Analysis

External calibration of XRF instruments is possible using matrix-matched standards to prepare a calibration curve. Sometimes, if suitable materials are not available commercially, this may require a sample matrix of very high purity, *e.g.* titanium dioxide, to be spiked with the oxides of analytes and then diluted with more titanium dioxide to the required concentration range for the analytes. This process obviously requires care when weighing the materials and to ensure thorough homogenization at all stages. Care should also be taken to match the standard matrix with that of the samples. For example, use high purity titanium dioxide for the standards if the samples under analysis are also titanium dioxide.

An external calibration curve is not always required for quantitative XRF analysis. Instead, an in-built response calibration can be used. This is pre-programmed into the instrument prior to supply. However, since both the detector and X-ray source efficiencies change over time, this in-built calibration should be monitored regularly and adjusted as and when necessary. For these ends, manufacturers usually supply a reference standard that may be run on an almost daily basis. The mean signal ± 2 SD may then be plotted for a suite of elements (typically 10–15). If, during the daily monitoring process, the signal of one or more elements is outside the acceptable range, then remedial action may be taken. It is then up to the analyst (and/or the agreed SOP of the institution) to decide what course of action to take. If one of the elements is outside the limits set but is not required for the analysis to be undertaken, they may decide that it does not matter as long as the analytes of interest are within the set limits. Alternatively, they may decide to take remedial action to try and ensure all elements are within the set limits. The remedial action to be taken would involve the analysis of some very well characterised materials also supplied with the instrument using a set of pre-programmed analysis protocols. The instrumental programme can then adjust the calibration accordingly. If an instrument suffers a mishap such as the deposition of a loose powder sample over the X-ray source or detector, taking remedial action is obligatory. It may take several calibration adjustment analyses to correct the signals.

Another problem with XRF analysis is the potential for inter-element interactions. These were discussed earlier. The use of certified reference materials matched as closely as possible to the samples under analysis will enable the analyst to estimate the magnitude of inter-element effects and possibly formulate methods for overcoming them.

A problem with the wavelength dispersive instruments is that they use a series of crystals on a goniometer that allows only a specific wavelength band of X-rays to pass through at any one moment. Once that wavelength band has been detected, the goniometer rotates, and another crystal allows a different wavelength band to pass through. Therefore, if numerous analytes need determination, a large series of crystal changes will be employed meaning that detection can be a very lengthy process. The number of crystals employed can potentially be decreased by picking analyte wavelengths that pass through the minimum number of crystals. However, this is likely to lead to decreases in sensitivity as "alternative", less efficient lines are chosen and may sometimes not be possible at all. Energy dispersive instruments do not have this array of crystals and are therefore capable of multiple analyte determination simultaneously but with some restrictions on resolution and selectivity.

3.5 Portable XRF Spectrometry

Portable XRF instruments have been around for many years. They started off being prone to errors because of difficulties in obtaining reliable calibration. This has become less of a problem more recently, but it should still be emphasised that the calibrations are usually inbuilt to the instruments and that different calibrations are required for different applications. For instance, when purchased, the manufacturer will programme the instrument with a series of calibrations that meet a particular laboratory's requirements. This could be for metal samples, environmental samples, thin films such as layers of paint, plastics, *etc.* Each of these applications has its own calibration (and set number of elements) and should ideally be sent for re-calibration/service every couple of years. Attempting to use a calibration for an application for which it is not designed is likely to lead to serious error. A readable tutorial on the use of hand-held XRF instruments was presented recently by Potts and Sargent.[4] An example of a portable instrument is shown in Figure 3.8.

3.5.1 Advantages and Disadvantages

There are many advantages associated with these instruments. The obvious one of these is that they are truly portable and may be used to analyse materials in- the field. This has the advantage of giving a geologist or environmental scientist an immediate answer rather than having to transport a sub-sample of material back to a laboratory. They are generally regarded as being completely non-destructive. This means that they are popular amongst archaeologists and conservators

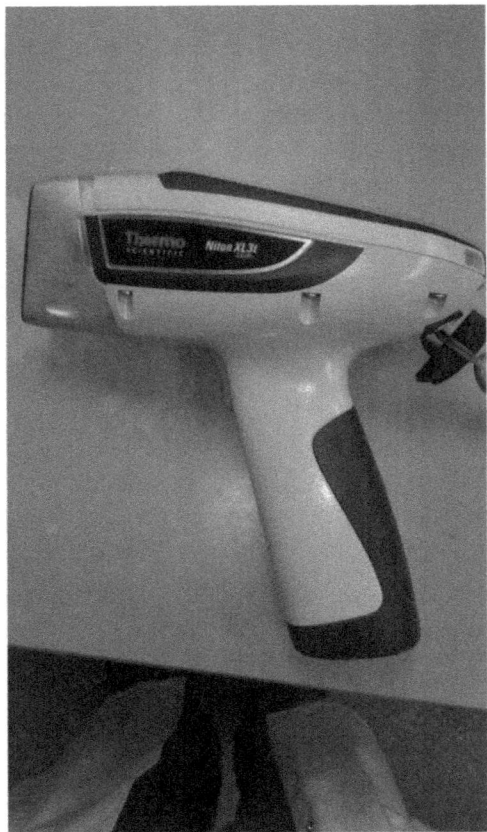

Figure 3.8   An example of a portable XRF instrument.

who are reluctant to destroy precious historical artefacts. They utilise energy dispersive detectors and so provide a result much more rapidly than wavelength dispersive, laboratory-based instruments. Most are equipped with a small camera and so it is possible to focus the X-ray beam onto a small area of sample. This is especially helpful when the X-ray beam may be decreased in size such that a circle of 3 mm diameter may be analysed rather than the usual 8 mm. This is of particular use when the sample under investigation is not homo-geneous, *e.g.* a painting with different colours.

The different built-in calibration types enable different parts of the same sample to be analysed. For instance, if an archaeological piece of pottery is to be analysed, the surface glaze may be analysed using the thin film calibration, whereas the main body may be analysed using the environmental calibration. Obviously, the usual precautions of removing surface contamination (*e.g.* soil) and not putting finger-prints on the sample are a requirement for obtaining reliable data. Similarly, the edge of the sample may have to be analysed to ensure that the glaze does not interfere with the results of the body of the ceramic.

One of this type of instruments' greatest assets is their ability to screen samples very quickly and easily. An example is the analysis of brominated materials in polymers. Rather than undertaking lengthy extraction procedures followed by chromatographic analysis of a large number of samples only to discover that 90% do not have any Br-containing compounds, it is much more efficient to undertake a swift portable XRF analysis to identify those that do contain them. The lengthy extraction procedures can then be undertaken on this much smaller group of samples, hence saving time, money and resources prior to the identification and quantification of the Br-compounds.

Another use for them is the analysis of loose powders. These can be prepared in sample cups similar to those described previously and analysed directly. Since the underside of the sample cell has the Mylar film and there is a second film usually placed across the orifice of the sample table to the XRF instrument, a double layer is formed, meaning that rupture is less likely. Even if rupture does occur, these instruments are far more easily cleaned and are less likely to receive serious damage if powder does land on the surface.

In this analyst's experience, these instruments tend to be really quite stable sensitivity-wise over a long period (assuming the ana-lytical surface has not been contaminated or damaged). This means that re-calibration by the manufacturer is not necessary for extended periods (often 2 years).

Battery life for these instruments can extend to several hours. However, if a replacement battery is required, they can simply be exchanged for the spent one and work can continue. The in-built memory on the instrument will have saved all the analyses under-taken and so those, plus the ones obtained using the new battery, can be downloaded to a computer at the end of the analysis session. It is therefore wise to have several batteries at your disposal so one can be used, one can be ready to replace it and a third can be re-charging.

The instruments may be triggered manually, *i.e.* held in the ana-lyst's hand, or it may be triggered automatically in a standoff fashion through the use of a laptop computer. The latter is obviously safer and is used more routinely. This is because the instrument is usually clamped in place using a rig or stand so that the same spot can be analysed for the entire duration of the analysis. This is not so easy to do when it is used in handheld mode. If the sample is homogeneous, then some movement in the beam location is not so much of an issue.

Among the disadvantages is the potential issue of safety. When used correctly, they are perfectly safe instruments. However, when used in the field it is possible to have an accident. They are designed not to activate if a target is not reasonably adjacent (this safety interlock should be tested periodically and the instrument returned to the manufacturer for repair if it fails). However, it is possible for an untrained user to carry the instrument in one hand with a finger on the trigger. In this way, it is possible to accidentally "shoot" yourself in the leg. Untrained users should therefore always be accompanied by experienced users until they are fully conversant with and adhering to the 'rules of operation'. Even in a laboratory environment, the instrument should not be left unattended. If the operator needs to leave the room, they should remove the battery and lock it away to ensure nobody can "play" with the instrument. A further safety requirement is that radiation warning signs should be placed near the instrument instructing other laboratory users to keep their distance. In general, the instrument should be regarded as a weapon – perfectly safe unless you do something stupid. A further safety precaution that can be observed if the instrument is not clipped into its holder is the wearing of a protective smock. This gives the operator protection over their body, but not their arms or their legs below the knees. These smocks are heavy, not terribly comfortable and, when worn when the temperature is high, they make the user very hot.

Another disadvantage of the instrument is its differing sensitivity depending on how far from the sample it is. In a laboratory setting, the instrument is usually clipped into a holder on the underside of a stand. Sample is placed on the upper surface of the stand, the shielded lid closed and the X-rays applied. If the samples are completely flat, they are all the same distance from the instrument and hence reliable results are usually obtained. For curved samples or samples with an uneven surface, this is not so easily achieved and so the analyst should be wary of the data produced. This can partially be overcome for the analysis of different parts of curved surfaces by moving the sample and then propping one end up so a part of the surface that was further away from the beam is now adjacent. Such a protocol can obviously not be employed for the inner surface though. A picture of a portable XRF instrument clipped into the underside of the stand is shown in Figure 3.9.

For some sample types, it is necessary to measure their depth using a digital micrometer, and this value is inserted to the instrumental

Figure 3.9 A picture of a portable XRF instrument clipped into its stand showing the surface on which samples are placed.

operating conditions. This is to ensure that the X-rays applied are not of sufficient power to penetrate the entire depth of the sample. The minimum depth required will depend on the material to be analysed, with less dense materials requiring a greater thickness to prevent the X-rays simply passing through.

The instruments tend to be less sensitive for the lower mass elements such as Al, Na, Si, *etc.* Sensitivity can be improved in two ways:

- the use of a polymer film to cover the orifice between the instrument and the sample that is especially designed for the determination of these analytes rather than the standard material and

- the use of a vacuum pump or a low flow of helium across the instrument's analytical surface to displace the air that otherwise absorbs the X-rays emitted.

Although the helium flow is very low, it can add to the expense of the analysis – especially if the analyst forgets to turn off the gas cylinder when they shut the instrument down. If they return to use the instrument a week later, they could discover that what had been a full gas cylinder worth several hundred pounds has emptied itself.

Although energy dispersive instruments are generally regarded as being slightly less accurate than wavelength dispersive ones, and portable instruments being less accurate still, they are still capable of providing good, accurate and reliable data. Their stability is a big part of this. Many archaeological studies insert data obtained using these instruments into chemometric packages such as principal component analysis. These packages look for patterns and associations in the data and can be used to determine if pieces of archaeological pottery are from the same area, or even made in the same kiln. This helps archaeologists to elucidate provenance, trade routes, *etc.* None of this would be possible if the instrument did not have that analytical stability.

3.5.2 Hints and Tips

Since these are energy dispersive instruments, the analytes are determined simultaneously and rapidly. If extra sensitivity is required, then the time of analysis can be extended. Although the sensitivity increase is not directly proportional to the measurement time, a significant improvement can be made by extending the measurement time from 30 seconds to 2 minutes. The drawback is reduced sample throughput, so most analysts make a compromise between sensitivity and throughput. Consequently, few analyses take longer than a few minutes.

The instruments will have a finite internal memory. Therefore, after a certain number of analyses have been undertaken, memory will be full. It will therefore be necessary to delete some or all of the data before further analysis can be undertaken. It is therefore wise to ensure that all the data required has been exported to a computer at regular intervals. This may just be the results (easily exportable into a spreadsheet programme, *e.g.* Excel) but may also be spectra. The number of scans storable can run into thousands, so it isn't necessarily something that needs to be done on a daily basis. However, if there are several users all using the instrument regularly, then it is annoying to try and locate all of them to ask what can and cannot be deleted.

Other hints and tips, *e.g.* purchasing several batteries, propping up one end of a sample with a curved surface so that different regions of it may be analysed more accurately, *etc.* are given in the advantages and disadvantages section.

3.6 Data Analysis for XRF Instrumentation

Data analysis for XRF instrumentation is relatively straightforward because, in many instances, virtually no calculations are required. If loose powders, larger fragments of a material or liquid samples have been analysed, then no calculations are usually required. If a pellet or fused glass disk has been prepared from ground and sieved samples, then care should be taken to ensure that the correct weights of the sample and weight of flux/wax binder are inserted to the instrumental software. For the pellets, typically 8 g of sample is mixed with 2 g of the wax. As long as the weights are close to these values, it is often not necessary to insert the exact weight. Therefore, if 8.0150 g and 1.9820 g are mixed, it is often sufficient to leave the values at 8 g and 2 g when programming the experiment since the effect on the final concentration is minimal. Similarly, if only 4 g of material is available, and it has been mixed with 1 g of wax, the proportion of 4 : 1 is maintained. However, it should be noted that if these weights differ significantly such that the ratio of 4 : 1 is not maintained, the exact weights used should be inserted. When a glass disk is analysed, only 0.9 g of sample is mixed with 9 g of flux. Since a large dilution is obtained, the exact weights of both flux and sample are inserted to the software. Typographical errors, *e.g.* inserting 0.9043 g of sample instead of 0.9943 g would have a significant effect on the concentrations of analytes calculated. A similar error of ~10% would result had 0.8943 g been typed into the software. When glass disks have been analysed, the loss on ignition should also be included in the calculations. **All of the analytes combined plus the loss on ignition should total 99–101% of the sample**. If it does not, then samples are likely to have to be re-prepared.

Poor precision when measuring replicate samples could indicate that some debris had contaminated the analytical surface on one or more of the replicate samples. This could just be a few dust particles up to and including a whole pellet that has fallen out of its aluminium cup and is sitting in the X-ray cavity. This would continue to inadvertently to be analysed for all samples until the error is discovered.

In general, most elements present at percentage levels are reported as their oxide, *i.e.* instead of giving a Mg concentration, it is given as MgO. A junior analyst should understand that a simple calculation would be required for direct comparison with the Mg concentration obtained using ICP–OES or most other techniques. In this case, if an MgO concentration of 2.4% was obtained, then the corresponding Mg concentration would be the relative atomic mass of Mg divided by the relative molecular mass of MgO, *i.e.* $24.3/40.3 \times 2.4 = 1.45\%$ Mg $(14\,500 \text{ mg kg}^{-1})$.

References

1. C. Vanhoof, J. R. Bacon, U. E. A. Fittschen and L. Vincze, *J. Anal. At. Spectrom.*, 2022, **37**, 1761–1775.
2. E. Margui, I. Queralt and E. de Almeida, *Chemosphere*, 2022, **303**, 135006.
3. Microsoft Word – Platinum Labware Care Guide (xrfscientific.com).
4. P. J. Potts and M. Sargent, *J. Anal. At. Spectrom.*, 2022, **37**, 1928–1947.

4 Atomic Absorption Spectrometry (AAS)

4.1 The Analytical Process

In truth, many of the errors, some of which are fatal for the analysis, occur well before the analyst gets close to using an analytical instrument. If the wrong sample or a sample not representative of the bulk is collected, there is nothing that even the most expensive instrumentation can do to remedy the situation. As discussed in Chapter 1, an example would include not running a tap for at least a minute before collecting a sample of tap water, because if the tap has not been run for an extensive period, metals will slowly leach from any metal pipework into the standing water. This "standing" water should therefore be flushed away and fresh water collected. It would be embarrassing to complain to a water authority that the water they are supplying is unsafe to drink if the water had not been flushed and hence most of the contamination arises through leaching mechanisms. Once the sample has been collected, the sample preparation procedure should be fit for purpose. The previous chapter on sample preparation should be consulted to ensure this enables the goal of the study to be achieved. Once this has been undertaken, the majority of analytical techniques require a calibration curve to be prepared. This is actually the instrumental response to a set of usually aqueous standard solutions prepared from a primary or stock standard. It should be borne in mind that for extra analytical robustness,

Practical and Technical Guides for Laboratory-based Chemists No. 1
Atomic Spectrometric Methods of Analysis
By Andrew Fisher
© Andrew Fisher 2025
Published by the Royal Society of Chemistry, www.rsc.org

calibration standards that are matrix matched with the samples should be prepared. This next section will describe errors associated with the preparation of standard solutions. It is appropriate to flame AAS, electrothermal AAS, ICP–OES and ICP–MS analyses.

4.1.1 Preparation of Standard Solutions

The first thing to bear in mind is that a new standard should always be purchased so that it arrives before it is needed, *i.e.* do not wait until an existing standard is finished before ordering a new one. If work in a laboratory temporarily comes to a halt for several days waiting for the new standard to arrive, then work-flow will clearly be disrupted, much to the ire of managers.

The next thing to consider is that just because a laboratory has paid money for a standard, it would be a huge error to assume that this standard is correct. This author has had instances where a supposed single element standard of 10 000 mg L^{-1} Mn was ~4500 mg L^{-1} and a quality control standard containing 100 mg L^{-1} of 26 analytes had one missing completely. An incoming standard should therefore always be checked for concordance with an existing one.

When preparing standards, the calculations should be clearly noted in the laboratory notebook and, ideally, checked by another analyst for errors. Glassware should be pre-cleaned and ready for use. The stock standards are a different class of standard that are both available commercially and traceable to a primary standard. These should be inspected to see if they are compatible. For instance, if both As and Ag are in the list of analytes and the 10 000 mg L^{-1} stock standards are in nitric acid (for Ag) and hydrochloric acid for As, then a mixture of both at 100 mg L^{-1} will lead to the Ag precipitating as the chloride. Under such circumstances, separate 100 mg L^{-1} standards should be prepared, and then, if a 1 mg L^{-1} standard is required, a volumetric flask should be half filled with diluent (the solution that is either matched to that in the samples, water or dilute acid) before the two analytes added. Precipitation effects are far less common or severe at much lower concentrations, and hence, a 1 mg L^{-1} standard is less likely to precipitate than higher concentrations.

However, at lower concentrations, some analytes may undergo a gradual loss process from solution over time, called 'plate-out'; where the analyte is 'lost' to the inside surface of glassware. This could potentially lead to an exponential-shaped calibration response rather than a linear one. Decreasing the pH of the diluent to < 4 should help minimise this. Many analytes suffer from this problem, especially at the $\mu g\,L^{-1}$ range.

Once the stock standards have been collected and the calculations made, then the pipettes to be used should be calibrated. This is usually performed on a 'daily before use' basis. In the schematic diagram presented in Figure 4.1, it is apparent that, preferably two adjustable-range pipettes need to be calibrated. The 10 000 $mg\,L^{-1}$ stock standard should be diluted first by taking 0.25 mL of it and then diluting to 25 mL, thus forming a 100 $mg\,L^{-1}$ standard. This should then be stoppered and mixed by inversion several times to ensure complete mixing. Then, 0.25 mL of this should be taken (using a different pipette tip) and diluted to 25 mL, *etc.*

It should be noted that all adjustable automatic pipettes, whatever their dispensing volume, are more accurate and more precise at the top of their range. It should also be noted that pipettes capable of dispensing smaller volumes tend to be less accurate and precise than larger ones. For instance, the tolerance on a pipette that can dispense volumes between 0.1 and 1.0 mL can be $\pm0.8\%$ at 1 mL, but 4% at 0.1 mL. A smaller pipette, *e.g.* one capable of dispensing volumes of 0.01–0.1 mL, may have an accuracy/precision of $\pm1\%$ at 0.1 mL, but be as poor as $\pm8\%$ at 0.01 mL and still be within manufacturer's specifications. Therefore, for best accuracy and precision (albeit at the cost of time of calibrating a second pipette), for the preparation of the 4 $\mu g\,L^{-1}$ standard in the scheme shown in Figure 4.1, it may be better to use a small pipette with a top volume of 100 μL rather than risk a potentially larger error using a larger volume pipette at the bottom of its range.

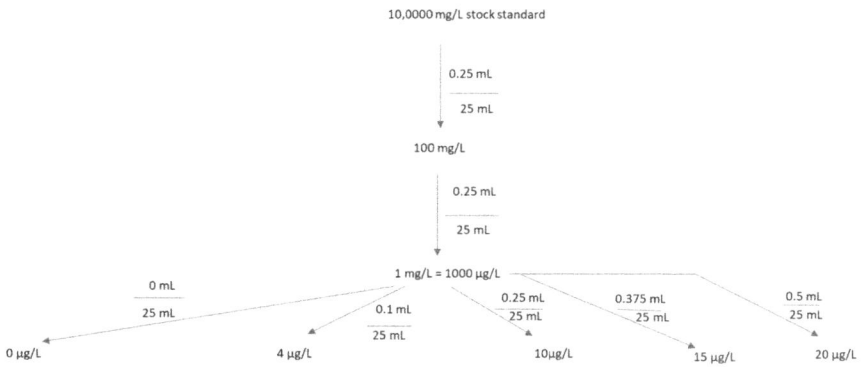

Figure 4.1 A schematic diagram for the preparation of some standards.

4.1.2   Calibrating a Pipette

Calibrating pipettes is an essential part of the preparation of standards. Consider the schematic diagram in Figure 4.1. If the pipette to be used is supposed to dispense 0.25 mL but only dispenses 0.225 mL, then a 10% error will occur. Another 10% error would occur for the next dilution step and again for the third. Under such circumstances, the calibration standards are appreciably below the nominal concentrations and hence, results for the samples will be significantly higher than they should be. Similarly (but less commonly), if the pipette dispenses higher volumes than expected, the concentrations of the standards will be significantly higher than the nominal values and hence, the data obtained for the samples will be too low. Such errors can be avoided by taking 10 minutes to calibrate the pipette(s) prior to use. Accredited laboratories will require evidence that a calibration has been made. Different laboratories will have different ways of accomplishing this. However, most will require the pipette code number (stamped into the pipette body), a date, the name of the analyst and the results of the calibration to be documented.

Before even touching a pipette there are several "do's and don'ts" that an analyst should obey.

Do:

- Pick the correct pipette for the volumes to be dispensed.
- Ensure that the correct pipette tip is placed on the barrel before use (ensuring the tip of the tip that enters liquid has not been touched by dirty fingers).
- Learn how the pipette works – each manufacturer has a slightly different mechanism, with some having locking systems to prevent accidental changing of the volume to be dispensed.

Do not:

- Adjust the pipette below or above its minimum and maximum values – this can break them if done at speed and with force so that the natural stop is violated.
- Place the pipette horizontally on a bench once the liquid is in the tip – it may flow into the barrel and damage the mechanism.
- If the tip cannot reach far enough into a vessel to collect the sample, do not invert it attempting to suck up the correct volume. This increases the risk of liquid entering the barrel and damaging the mechanism.

Pipettes should be calibrated using a calibrated balance. The balance should therefore be calibrated first with a series of 'certified' calibrated weights that are appropriate for the masses to be weighed, *i.e.* there is no point in calibrating with 25 g, 10 g, 5 g and 1 g weights if the masses to be weighed are 0.01–0.5 g. Sets of calibration weights can be expensive and should be handled carefully. Depending on the range of the weights, they will be supplied with a certificate of their masses and either a pair of forceps which should be used to pick them up (never use un-gloved fingers) or a glove. Some of the smaller weights (*e.g.* 10 mg) are fiddly to pick up with tweezers. Real care should be taken not to drop these on the floor and then damaging them further by accidentally treading on them. The weights should be re-calibrated at least once every 2 years by a suitably qualified facility. Accredited laboratories will require evidence that the balance has been calibrated before use, so the masses determined should be documented appropriately.

Once the balance has been demonstrated to be working within specifications, the pipettes may be calibrated using a weighing boat and pure water. Typically, a pipette will be calibrated at three points over its range, *e.g.* for a 0.1–1 mL pipette, three replicates at each of the volumes 0.1 mL, 0.5 mL and 1 mL should be made. If the minimum volume to be dispensed during an experiment is going to be 0.25 mL, then this could potentially be the lowest volume at which it is calibrated. The density of pure water is approximately 1, so therefore 0.25 mL should weigh close to 0.25 g. This is dependent on both temperature and pressure, so may deviate marginally. A typical weight for 1 mL will be in the range of 0.99700–1.0050 g. Many laboratories have a little table of acceptable weight ranges for different volumes dispensed printed out and placed close to the balance bench for easy reference. It should certainly not be assumed that if a pipette dispenses the correct volume at 0.25 and 0.5 mL then it will continue to do so at 1 mL. These mechanical pipettes can and do go wrong fairly regularly and require regular maintenance to work adequately.

It should be noted that as soon as the liquid is dispensed onto the weighing boat, it will begin to evaporate, *i.e.* the weight will start to decrease. The rate at which it does this is negligible for large pipettes, *e.g.* when dispensing a 5 mL volume. However, if attempting to calibrate using 0.025 mL, the evaporation rate can be significant as a proportion of the mass to be dispensed and hence the reading should be taken reasonably promptly. Capped or sealed weighing vessels may be used for weighing volatile liquids.

When pipetting large volumes onto a weighing boat (or a beaker), then do not do it with a flourish. Expelling the liquid at speed can make it squirt with force and potentially splash out of the receptacle and onto the balance itself. If it then enters the mechanism of the balance, it may take a while to dry and delay subsequent work.

There are two things to remember when using stock standards. It is a **'cardinal sin' to put a pipette directly into a stock standard bottle** as it may contaminate it. Instead, a small volume may be dispensed into a clearly labelled centrifuge tube or other container and then the pipette tip inserted to that. Any excess may be kept and used in the future. **It should not be replaced in the stock bottle.** Lids of stock bottles should not be left off for any longer than necessary, otherwise, evaporation will occur, hence artificially increasing the concentration of the analyte. Stock standard lids should also not be used as receptacles into which the standard is poured ready to be pipetted. This is because remnants may dry and so when the next time some is dispensed into the lid, a change in concentration occurs as the $10\,000$ mg L^{-1} stock re-dissolves the now solid material.

It is a good idea when preparing a calibration curve to make another "check" standard from an alternative stock solution. Clearly, in the scheme shown in Figure 4.1, a check standard of 10 µg L^{-1} would be appropriate. This should be prepared from a completely different stock standard – either from a different manufacturer or from the same manufacturer but from a different batch number. This will help identify if one of the stock standards has become contaminated. Accredited laboratories will require documented evidence of the serial or batch numbers of the two different stocks.

Standards are not stable for ever. The lower the concentration range, the less time they will remain fit for purpose. Standards of <10 µg L^{-1} should probably be prepared fresh every day. However, intermediate stocks, *e.g.* the 1 mg L^{-1} stock in the scheme above, are probably stable for several days/weeks, although this would be dependent on numerous factors including the analyte, the solution matrix (water or acid), the salts present and their concentration.

4.2 Flame AAS

Flame AAS is one of the simpler analytical techniques and has been available commercially since the 1960s but was devised in the mid-1950s. It has historically been regarded as being a single element technique, *i.e.* it is capable of determining only one element at a time.

However, more recently, multi-element instruments have been available commercially.

The theory behind atomic absorption spectrometry is covered in another volume of this series. For a basic introduction, the book by Ebdon *et al.* would be a good place to start.[1] The book is over 20 years old now, so the most recent developments, *e.g.* the simultaneous instruments, are not covered. However, the basics of how atomic absorption occurs, the origin of alternative wavelengths, interferences and methods of overcoming them, flame processes, *etc.* are all discussed. A very brief introduction to the theory is required to aid the reader. Normally, a liquid sample is aspirated into a flame which dries and atomizes the sample, *i.e.* analyte molecules are dissociated into atoms. Light from a source that is usually a hollow cathode lamp, electrodeless discharge lamp or, for multi-element instruments, a high intensity continuum source is shone through the flame. For single element instruments (*i.e.* instruments that can only detect one element at a time), light characteristic of the element coating the cathode is emitted. Therefore, if the lamp is a magnesium hollow cathode lamp, light characteristic of magnesium is emitted. This light is of exactly the correct wavelength to be absorbed by magnesium atoms that have been formed in the flame. The amount of light absorbed is therefore proportional to the concentration of magnesium in the sample.

4.2.1 Advantages and Disadvantages of Flame AAS

The advantages and disadvantages of flame AAS analyses are summarised in Table 4.1 and then discussed in slightly more detail in the text.

Table 4.1 Advantages and disadvantages of flame AAS.

Advantages	Disadvantages
Cheap to purchase	Requires the use of explosive gases
Very few spectral interferences	Relatively poor sensitivity (analyte dependent)
Very rapid per measurement	Short linear range
Can easily be coupled with alternative sample introduction systems.	Can consume significant volumes of sample
Easily automated	Requires sample introduction as a liquid (or gas)
	Relatively more non-spectral and chemical interferences compared with ICP–OES

4.2.1.1 Advantages

A basic instrument is actually relatively cheap, although the more complex optics required for the multi-element instruments makes them significantly more expensive. They would still be cheaper than an ICP–OES instrument and significantly cheaper than an ICP–MS instrument though. For the cheaper instruments capable of determining only one analyte at a time, hollow cathode lamps or electrodeless discharge lamps for each of the analytes of interest are required. These can cost well over £100 each; much more so for some analytes and for lamps with more than one analyte. Flame AAS systems are also quite cheap to run, using acetylene (typically) and often air provided by a compressor. For some analytes, *e.g.* refractory elements such as Ti, Si, W and those that may form stable compounds, *e.g.* Ca, a hotter flame will be required. This often means using nitrous oxide instead of air as the oxidant. Although this is not cheap, a cylinder can last a long time (often months).

One of the biggest advantages of AAS is that it suffers very few spectral interferences. This is because of the "lock and key" effect discussed in other volumes of this series. Assuming the correct monochromator bandpass is used, then very few truly "spectral interferences" exist. The analyst should not take this to mean that interferences as a whole do not exist. Chemical interferences formed within the flame system can seriously affect the sensitivity of an analysis. For instance, if Ca is to be determined in a fertiliser sample that has considerable amounts of phosphate present, the refractory compound calcium phosphate may well form which will not be atomized as efficiently as calcium nitrate present in the standard solutions. A severe underestimate of the Ca content of the fertiliser would result. Matrix matching or the use of the standard additions technique can help overcome this problem. Alternatively, the addition of a releasing agent, *e.g.* a lanthanum salt, that will form a phosphate in preference to the calcium phosphate, will release the Ca atoms to be determined more accurately. The use of the much hotter nitrous oxide – acetylene flame will also be beneficial since it produces a much higher atomization efficiency than a standard air-acetylene flame. Another potential problem is that of easily ionized elements. The formation of ions is far from ideal because they do not absorb light at the same wavelength as the atoms. These easily ionized analytes can cause problems because the presence of other easily ionized elements in the samples will contribute electrons, forcing many of the analyte ions back to the atomic form. If these concomitants are not present in the standards, the analytes remain in the ionic form. It is therefore possible for a pure standard of an analyte to have the same concentration as in a sample, but to give a completely different (lower) signal. Under such circumstances, large over-estimates of that analyte's concentration in the samples will result. This problem can be overcome by adding an excess of an element (but not an analyte) that is more easily ionized than the analyte of interest to both the sample and standard solutions. This problem exists mainly for the alkali and alkaline earth elements but can occur for others if the hot nitrous oxide – acetylene flame system is applied. If K and Na are both analytes, then the addition of Cs to both standard and sample solutions would overcome easily ionized element effects. Ideally, the ionization suppression element should have been more easily ionized than the analytes. Therefore, Cs would be used in preference to Li.

The instrument manufacturers usually provide a handbook. This contains wavelengths available for each analyte, the monochromator bandpass (also erroneously called the slit width) for each wavelength, the flame-type required and potential interferences. These handbooks are invaluable and should be consulted if a new analyte is to be determined or a different sample type is to be analysed.

From the moment, a sample is introduced to the nebuliser uptake tube to taking measurements may take only 3 or 4 seconds. It may take another few seconds to take replicate readings. The measurement stage is therefore extremely quick. A little patience should be exercised to wait those first 3 or 4 seconds for the sample to reach the flame. If replicate readings are taken too quickly, the first replicate may have a significantly lower absorbance (and hence effective concentration) than the others – an indication that the sample solution had not reached the flame over the entire integration period and the signal had not yet stabilised. If the analyst does not notice the first replicate is significantly lower than the others, the mean result may be skewed and a poorer standard deviation will result.

Another advantage of flame AAS is that it may readily be coupled with other sample introduction techniques or devices. Hydride/vapour generation, flow injection techniques, chromatography, *etc.* may all be coupled with minimal effort or inconvenience. The natural uptake rate of a flame AAS nebuliser is typically 5–8 mL min^{-1}. Therefore, if a chromatography technique that has a flow rate of 1 mL min^{-1} is to be coupled with it, a small air bleed may be necessary in the coupling to make up for the difference in flow rate.

4.2.1.2 Disadvantages

One of the main problems associated with Flame AAS is the use of combustible and hence potentially explosive gases, *i.e.* acetylene. Many countries have brought in health and safety regulations prohibiting acetylene cylinders from being brought into laboratories. Instead, they must be kept in a cage adjacent to the building for easier access for the fire and rescue services. The added cost of piping the acetylene into the building and sometimes up many stories, having it leak checked every year, *etc.*, can be off-putting. In addition, flash arrestors should also be purchased and installed to minimize the risk of flames going through the pipes into the cylinders.

The sensitivity and limits of detection (LOD) of flame AAS instruments are relatively poor, with LOD being typically at the tens of $\mu g\,L^{-1}$ for many analytes and significantly poorer for some refractory elements (even if a hot nitrous oxide – acetylene flame is used). This is a function of the nebulisation efficiency being only 10–15%, *i.e.* if 1 mL of sample is nebulised into an instrument, only 0.1–0.15 mL will reach the flame. The rest is discriminated against by the spray chamber and directed to waste. Another contributing factor to the poor sensitivity is the residence time of the analytes in the light beam. When a sample enters the base of a flame it travels in an upward direction through the flame quickly and passes through a 0.5 cm beam of light produced by the hollow cathode lamp in significantly less than a second. Greater absorbance (and hence better sensitivity and lower LOD) would be achieved if its passage is impeded so that it stays longer in the light beam. Methods for achieving this will be discussed in the electrothermal AAS section of this chapter and the hydride/vapour generation chapter. It should be noted that each element will have a series of different wavelengths available to use analytically. Each of these different wavelengths will have its own linear calibration range, its own LOD, *etc.* The linear calibration range is that part of the calibration curve that is linear rather than curved and is the part from which analyses should be made if the most accurate data are to be obtained. A third contributing factor may be the incomplete atomization of some refractory compounds. For some analytes, even the use of the hotter nitrous oxide – acetylene flame may be insufficient to break up compounds into their constituent atoms efficiently. An example is silicon in the form of silicates. This is very stable and is incompletely atomized – even when the hotter nitrous oxide – acetylene flame is used. Consequently, the LOD for Si is closer to $1\ mg\,L^{-1}$. For some elements, the opposite is true. Their compounds are very easily split and the resulting atoms are easily ionized. This phenomenon was discussed at length previously.

Since flame AAS is an absorbance-based technique, it will have a very limited linear range, spanning probably only two or three orders of magnitude. This compares poorly with emission, fluorescence and mass-spectrometry-based techniques where linear ranges can easily reach five orders of magnitude and sometimes more. As discussed in the previous paragraph, each wavelength of an analyte will have its own linear range, LOD, *etc.* For instance, Cu at the main 324.8 nm line will have an LOD of $\sim$20 $\mu g\,L^{-1}$ and a calibration curve that will be linear up to approximately 5 $mg\,L^{-1}$. Poorer LOD, but linear ranges extending beyond this are possible if an alternative wavelength is used. For instance, the 327.4 nm line will have an LOD of $\sim$0.1 $mg\,L^{-1}$ and a linear range extending to over 20 $mg\,L^{-1}$.

As discussed previously, the natural sample uptake of a flame AAS nebuliser is often in the range of 5–8 $mL\,min^{-1}$. It may take only 10 or 15 seconds for a single determination of an analyte to be made, so this may consume $\sim$1 mL of sample. This could potentially be problematic if a very limited sample volume is available, *e.g.* 5 mL and if 10 analytes need to be determined. Under such circumstances, it is possible to run out of sample before all analytes have been measured. There are ways around this problem. For instance, a discrete sampling cup (a Delves type cup) could be attached to the nebuliser uptake tube. The nebuliser tube attaches to an orifice at the bottom of the cup. A discrete volume of the sample could then be injected using a pipette (*e.g.* 0.1 mL) into the cup and the nebuliser passes it to the flame in the normal way. The sample absorbance would be observed as a transient signal. If the integration time of the instrument is set appropriately (*e.g.* for 3 or 4 seconds), the area under the whole transient can be measured accurately and precisely. In this way, three or four replicates of the same sample may be made using a total volume of <0.5 mL per element. This procedure is clearly not as straightforward as conventional sample introduction, but is a good way of conserving sample and, with only a relatively small amount of work, can be optimized in terms of time and sample volume consumed.

For most flame AAS analyte measurements, the sample needs to be introduced as a liquid or, in the case of hydride generation sample introduction, as a gas. This means that if the sample type is a solid, one or more of the sample preparation methods described in Chapter 2 are necessary for the analysis to occur.

4.2.2 Hints and Tips

As with all types of analysis, it helps to know what is in your sample before you start the analysis. For some sample types, *e.g.* for quality control purposes in a foundry or a factory, this may well be the case. Clearly, this cannot always be the case though. However, sometimes a "best guess" approach can be useful. For instance, if a series of environmental samples, *e.g.* 100 sediment digests are to be analysed for the Fe content, then it is useful to know that typical Fe concentrations are likely to be in the range of 10 000–50 000 mg kg^{-1} in the solid material. Therefore, after 1 g of material is digested and then diluted to 100 mL volume, the concentrations in the digests are likely to be somewhere in the range of 100–500 mg L^{-1}. This is not always the case (*e.g.* if the sediment is high in sand content), but as a general "rule of thumb," this range is appropriate. The analyst therefore has a choice to make. They could:

- use the most sensitive Fe wavelength at 248.3 nm, where the linear range extends only to 5 or 10 mg L^{-1}, making calibration standards in that range and then diluting the sample digests by a factor of 100 (think of the time and materials required to do this!) or,
- they could use the much less sensitive Fe wavelength of 392.0 nm that has a linear range extending to over 2000 mg L^{-1}, making appropriate calibration standards for this and then running the large majority of the samples directly.

Although in many incidences an alternative, less sensitive wavelength may be used, this is not always the case. For instance, the only alternative for the main Cd wavelength at 228.8 nm is 326.1 nm which is over 300 times less sensitive. It is most unlikely that there are too many samples with sufficient Cd content for that to be of use – the possible exceptions being if an analyst is checking the Cd content of some cadmium telluride nanodots or is analysing a battery or cadmium-based pigment.

Under certain circumstances, it is possible to extend the linear range of a wavelength. If the burner head is rotated slightly, then the light beam from the hollow cathode lamp no longer passes through the full length of the flame. This shortened path length means that from the Beer–Lambert Law, higher concentrations may be measured. This is particularly useful when it is suspected that most samples fit onto the calibration range, but a few are likely to be slightly above. If, for instance, most Zn results in a sediment digest are likely to be <1 mg L^{-1}

(the top of the linear range for the most sensitive wavelength of 213.9 nm), but a few are likely to be 1.1 or 1.2 mg L^{-1}, then the burner head can be rotated, calibration standards of 0.25, 0.50, 0.75, 1.0 and 1.25 mg L^{-1} can be prepared, checked that they are linear and then the samples run directly. An alternative to this is diluting those samples that are just over the 1 mg L^{-1} top standard. Again, the choice would be the analyst's, but if most samples require a small dilution, it may be quicker and easier to rotate the burner head. **Note: the burner head should be rotated before the calibration standards are run, because the signal for the standards will not be the same before and after the rotation.**

An alternative to rotating the burner head would be to fit the burner used for nitrous oxide – acetylene. These tend to have a slot length half that of those used for air – acetylene and hence would have half the sensitivity. **Note that the converse may not be possible. Sensitivity doubles with increased path length, but if a nitrous oxide – acetylene flame is to be used, safety interlocks of the instrument should prevent the longer burner head for air-acetylene from being used since it is likely to lead to an explosion.**

Each analyte will have an optimal flame chemistry. Even those that require an air-acetylene flame may require either a fuel lean (hot and oxidising) flame, a fuel rich (cooler but very reducing flame) or something in the middle (a so-called stoichiometric flame). The instrument manufacturer's handbook should give the analyst the gas flow settings for the correct flame type and flame chemistry. However, it is important to note that some sample types alter the flame chemistry significantly. For instance, alcoholic spirits contain ~40% ethanol. This will have a very big effect on the flame chemistry. Other "organic" samples *e.g.* dichloromethane extracts of petrochemicals, APDC/IBMK extracts of water samples, *etc.* will have a similar effect. This should be taken into account when optimising the flame chemistry for optimal sensitivity.

Following on from that last point, organic-based solvents will have a very different vapour pressure and surface tension to aqueous-based solvents. Therefore, the transport efficiency to the flame is likely to be greater than the typical 10–15% observed for water or dilute acids. This means that standard solutions matrix matched to those of the samples are required to prevent catastrophic errors. This would also have an effect on the LOD, sensitivity and linear range and so should be considered prior to analysis.

The use of organic solvents may lead to another problem. After extended use, small amounts of carbon can be deposited and may begin to clog the burner head. This leads to "flame-voids or holes" appearing in the flame, effectively decreasing the path length and decreasing the sensitivity. These particles can be removed (usually) using a spatula or other non-combustible material. It is therefore advisable to check after every tenth sample with a standard to ensure signal stability. The same problem of carbon deposition is noted with the nitrous oxide-acetylene flame, even for aqueous-based samples. This is because this flame uses much larger volumes of acetylene than the air-acetylene flame. A very fuel-rich air-acetylene flame may also lead to the same problem.

Analysing a standard every 10 samples also helps detect other errors. If particulates are present in sample digests, they can block the nebuliser uptake tube or the nebuliser itself. If this occurs, then clearly either no signal will be observed or, if it is a partial blockage, a much smaller signal results. Blockages may be cleared using a nebuliser wire. However, the holes involved in the fine bore uptake tubes can be very small and decent eyesight helps enormously. If a blockage has been removed, it is necessary to either re-calibrate or, at the very least, use a check standard to ensure there has not been a change in sensitivity. It is also worthwhile ensuring that the absorbance signal returns to 0 between each sample. If it does not, then a shift in the baseline will cause errors during the calculations. This is especially problematic if determinations close to the LOD are being performed.

When first starting a flame AAS instrument on any day, it is worthwhile making a few checks before turning on the flame and starting an analysis. It is worth checking that the light beam passes the full length of the burner slot. If a previous analyst has rotated the head and not returned it to the usual position, then a decrease in sensitivity would result. This checking may be achieved using a piece of card and moving it from one end of the burner to the other whilst ensuring that the light beam passes along the length of the burner slot. **Note: ensure the flame is not lit while doing this**. It is also worth checking that the correct burner head is fitted. As discussed previously, a nitrous-oxide burner head has a much smaller slot length than one for an air acetylene flame. Modern instruments will not allow a nitrous oxide – acetylene flame to be lit if the incorrect burner head is fitted because of the risk of explosion. As a side note, the safeguards on modern instruments are far superior to those of 40 years ago, when the safety precaution for using nitrous oxide was to chain the burner to the instrument chassis so that it did not embed itself in the ceiling. Before starting an analysis, it is worth turning the instrument on and allowing the main electrical system (including the hollow cathode lamp) to "warm up" for a period of at least 20 minutes before any measurements are made. This will help minimize signal drift. It is also worth checking the condition of the hollow cathode lamp. If the cathode coating is now heavily coating the glass sides and the discharge is flickering, the lamp has probably reached the end of its useful life. Attempting to use it will likely lead to a very noisy signal and make it almost impossible to "zero" the instrument on a blank or the wash water. Many modern instruments will also have an error message for this. Checking that the water trap contains sufficient water is also a necessity. This is to prevent any unburnt acetylene from entering the laboratory. Most instruments will have a safety interlock for this. Similarly, all fume extraction hoods and oxidant gases should also be turned on prior to use. Again, if anything has been forgotten, the instrument's safety interlocks should prevent disaster.

Modern instruments will plot the calibration curve and then take concentration readings direct from this ready for export. It is usually still possible for the analyst to use the absorbance values and plot the graphs manually and then calculate the concentrations off-line. Both methods have advantages. Allowing the instrument to do all of the calculations is clearly less labour intensive. However, if the analyst makes a typing error in the weights of samples used, it will need a return to the instrument to correct it. Doing it all off-line is more labour intensive and time-consuming but any errors are more easily identified and corrected. If one allows the instrument to do all calculations it is therefore worthwhile double checking all the relevant weights, volumes, *etc.* whilst programming the instrument.

For multi-elemental instruments, a continuum source is used rather than hollow cathode lamps (or electrodeless discharge lamps). Instead of a relatively simple Czerny–Turner monochromator, they use a much more complex Echelle type spectrometer. Their simultaneous nature clearly overcomes the problems of being time-consuming and the consumption of excessive volumes of samples. It should be noted though that if analytes are to be determined simultaneously, the flame chemistry needs to be the same for all analytes. This may require a compromise set of conditions to be used that are ideal for none of the analytes, but adequate for all. This may mean there is a concomitant drop in sensitivity and a poorer set of LOD being obtained.

Overall, when using flame AAS, it should be remembered that although an individual reading may be fast, if 10 analytes require determination, this could potentially require 10 different dilution steps to ensure the analyte is on the linear part of each analyte's calibration curve. Therefore, individually they may be determined at speed, but collectively, the amount of time required is significantly longer than a multi-elemental technique such as ICP–OES. This previous statement does not apply to the simultaneous instruments. In truth, the sales of AAS instruments have been in decline for years as laboratories have switched to the multi-elemental techniques, *e.g.* ICP–OES. However, the simultaneous AAS instruments may extend the technique's application. It is worth noting though that if a small company, *e.g.* an electro-plating company, needs to determine only one analyte in their waste sludge to ensure that it complies with local waste disposal rules, then a basic AAS instrument is among the cheapest and most efficient options for the analysis.

4.3 Electrothermal AAS

Commercial instruments have been around since the early 1960s and it was the "ultra-trace" technique of choice for a further 30 years. Then, the quasi-simultaneous ICP–MS technology was developed that had similar LOD and slowly, the number of ETAAS instruments in circulation has declined. Again, simultaneous versions that enable more than one analyte to be determined at once are available, but it is possible that over the next decade or so, they will slowly start to be phased out. A readable account of the advances over the last 30 years in simultaneous instrumentation and applications was provided recently by Pasias *et al.*[2]

The theory of ETAAS (also known as graphite furnace (GF)-AAS or electrothermal vaporization AAS (ETVAAS)) is given in a different volume of this series. Briefly though, a sample, usually in the form of a liquid, is injected into an electrically conducting graphite tube and the tube then undergoes a program of increasing temperatures to:

- dry the sample;
- ash/char the sample to remove as many concomitant species as possible in an attempt to decrease interferences;
- atomize and then vaporize the analyte of interest into the light beam from the hollow cathode lamp and finally
- clean any remnants of the sample prior to the next sample or sample replicate being injected.

The temperatures of each of these stages depend on the analyte and the sample type. The drying stage is often a multi-stage affair with temperatures above the boiling point of the sample solvent. For instance, water samples could be dried for 30 s at 110 °C and then for a further 20 s at 140 °C. A nitric acid digest of a material may require these temperatures to be modified to perhaps 130 °C followed by 160 °C. The drying temperature should be sufficiently high to allow the solvent to evaporate slowly and gently. Too high a drying temperature may cause the sample to spit or froth. This may lead to analyte loss through it escaping from the tube or poorer precision. Analysis of proteinaceous materials such as blood can require extensive time periods to dry very slowly. This is because it does not evaporate to dryness in the same way as a water sample. Instead, it "cooks" a bit like an egg, where the outside of the sample droplet solidifies and then the middle slowly does the same. If too high a temperature is used, the pressure within the drying sample droplet can increase to the extent it "explodes" and splatters around and possibly out of the tube. The result would again be a potential loss of analyte or extremely poor precision. The ash/char stage is where the concomitant elements and the matrix material are removed. Different analytes have different melting/boiling points and so the temperature used depends very much on the analyte of interest. In general, the highest temperature that can be used without analyte loss is chosen to maximise interference removal. This temperature is usually between 350 and 1200 °C with a duration of perhaps 20–30 s. This stage will be discussed in greater length later. The atomization stage vaporizes the analyte into the light beam for measurement. This will again be analyte-dependent, but often temperatures between 1600 and 3000 °C are required for a time period of 5–7 s. The clean stage is usually at maximum power, *i.e.* 2800–3000 °C for a further 5 s.

4.3.1   Advantages and Disadvantages of ETAAS

The advantages and disadvantages of ETAAS are summarised in Table 4.2 and will be discussed at greater length in the text.

In general, ETAAS shares many of the advantages and disadvantages with the flame version, but there are important differences,

4.3.1.1   Advantages

They are far from cheap, but a basic instrument costs significantly less to purchase than an ICP–OES instrument. A simultaneous instrument's cost to purchase will be comparable to an ICP–OES instrument. Running costs are also relatively low. They use argon to

Table 4.2 Advantages and disadvantages of ETAAS.

Advantages	Disadvantages
Cost of purchase is less than other instrument types yielding similar sensitivity.	Slow analysis times
High sensitivity and low LOD	Short linear range
Consumes very little sample	Prone to significant interferences
Easily automated	Some limitations to its simultaneous multi-element capabilities
May analyse solid or liquid (or gaseous) samples	

prevent oxidation of the graphite atom cell, but the amount used is significantly lower than that by plasma-based instruments. The other running costs include the light sources (hollow cathode lamps or electrodeless discharge lamps) and the graphite tubes used for the atom cell. These tubes can be fairly expensive, depending on the design, the material used (electrographite or pyrolytic graphite), *etc.* Their lifetime depends on what analyses they are used for. A tube used to determine Cd in river water samples will last a very long time because the temperatures used will be low and the sample is not chemically aggressive. However, if the analyte is Si, this requires much higher temperatures for it to atomize, hence tube lifetime will be diminished. If the sample type is chemically aggressive, *e.g.* contains very high concentrations of acids or oxidising agents that attack the graphite surface, the tube life time will be diminished further.

The detection limits are very low for ETAAS, with values ranging between <0.05 $\mu g\,L^{-1}$ for the extremely sensitive analytes, *e.g.* Cd and Mg up to 10 $\mu g\,L^{-1}$ for analytes that form stable carbides, *e.g.* W. The huge improvement in LOD compared with the flame version of AAS is because of two main reasons. The sample is pipetted directly into the atom cell, *i.e.* sample transport is 100% compared with 10–15% in FAAS. In addition, the light beam passes through the tubular atom cell. As the analyte atoms are vaporized into the light beam, they may take 2 or 3 seconds to flow out of the tube. Their residence time in the light beam is therefore greatly increased compared with FAAS. The longer an atom spends in the light beam, the more likely it is to absorb the light hence, greater absorbance. For ETAAS, the detection limit is often given as an absolute amount rather than a concentration. For instance, if the LOD for Cd is 0.05 $\mu g\,L^{-1}$ and a 25 μL aliquot of sample (or standard) is injected into the atomizer, the absolute LOD is 1.25 pg.

As can be seen from above, a very small volume of sample or standard (10–50 µL) is injected into the instrument. Therefore, it consumes very little of the sample material. It should be noted of course that replicate readings need to be made, but the overall volume consumed per analyte is still less than 0.25 mL and often <0.1 mL. For simultaneous instruments, several analytes may be determined in these small volumes.

Samples can be introduced using an automatic pipette. When manually injected, this is obviously very labour intensive and actually leads to poorer precision for replicate readings. This is because the position onto which the sample is introduced is critical. An auto-sampler is therefore used for most analyses, since this will introduce the sample to the same location every time. Modern autosamplers are much more than simple sample delivery systems. It may introduce the sample, dilute it if it is found to be too concentrated for the cali-bration curve, add matrix modifiers (more about these later), and they can even prepare standards by diluting one standard presented to it. For instance, the analyst can prepare a standard of 20 $\mu g\, L^{-1}$, place it in an autosampler cup, place some diluent in another autosampler cup and instruct the autosampler to construct a calibration curve of 0, 5, 10, 15 and 20 $\mu g\, L^{-1}$ from it. Many analysts still prefer to prepare their own standards though. The autosampler is driven through the instrument's computer and can be instructed to perform QC checks, *i.e.* to run a standard every 10 samples. If the QC check is out of the tolerance limits input to the computer, the autosampler can be in-structed to:

- re-calibrate and continue the sequence of samples,
- re-calibrate, re-run those samples since the last good QC check and then continue the sequence of samples,
- continue the sequence of analysis anyway ignoring the fact that something is wrong,
- stop the analysis and wait for an analyst to sort out whatever the problem is.

The autosampler is therefore a very efficient tool, freeing up ana-lysts to do other tasks whilst providing data that is probably of better quality than the analyst could obtain manually.

Unlike flame AAS instruments, ETAAS instruments can analyse solid samples directly. This normally requires the use of specialised graphite tubes with slots machined into the sides through which little platforms with solid sample material can be placed. Such analyses cannot be automated easily. They also suffer from poor precision because analysing just 1 or 2 mg of material can be difficult in terms of obtaining an accurate weight, ensuring that absolutely none of the material weighed onto the platform is lost during transfer to the graphite tube, and ensuring that the very small sample weight transferred is a reproducible, representative fraction of the bulk sample each time. The huge majority of analyses undertaken therefore do not use the solid sampling methodology. Numerous papers have been published that introduce solids as slurries though. This negates the necessity of complex sample preparation methods, but does require a homogeneous mixture, *i.e.* a dispersant should be used to prevent particles flocculating together.

4.3.1.2 Disadvantages

As discussed above, a typical program requires the sample to be dried, ashed/charred and atomized. Then, the tube is cleaned prior to the next sample injection. The duration of each of these stages varies, but overall, the majority of ETAAS analyses take between 1.5 and 2.5 minutes per replicate (plus the time required for the autosampler to collect the sample and matrix modifier and then dispense it into the graphite tube). If three replicates are made per sample, then the time required can be up to 8 minutes per sample. This is obviously very unfavourable compared with other techniques. Remember too that the majority of instruments may only determine one analyte at a time. The simultaneous instruments mitigate this problem slightly by determining several analytes in the time it takes to determine one using a more basic instrument. However, this would entail running compromise conditions for the tube, thereby limiting some of the single analyte advantages of performance and interference. This is because the temperature program used must be appropriate for all of the analytes. There is no point in using an ash temperature of 1000 °C because it is the optimum for Mg and Mn if the one of the other analytes is Cd, because this will be lost at such an elevated temperature. Some research papers have utilised "hot injection". This helps reduce the time required per sample run to potentially only a minute or so, but it has not found widespread popularity among the general analytical community who have preferred to maintain rather than develop new standard operating procedures (SOPs).

In common with flame AAS, ETAAS is still an absorbance-based technique. It therefore has a relatively short linear calibration range spanning only two or three orders of magnitude. Also in common with FAAS, alternative wavelengths may be chosen for most analytes, so a dilution is not always necessary. This is especially useful for the simultaneous instruments, where if three analytes have concentrations that are on the calibration curve but a fourth is just too high, the samples would have to be re-run with dilution just for that fourth analyte. By picking less sensitive wavelengths, time and money can be saved. Since it may not be known from the outset that the fourth analyte will be just too concentrated for the wavelength chosen, it may take a little effort to identify the concentration range required and then choose the most relevant alternative wavelength. This need not be terribly time-consuming and, once set up, can be used routinely for the analysis of further batches of the same sample type. Alternatively, multiple wavelengths may be chosen for any analyte and an extra standard programmed in to be used just by the less sensitive wavelengths.

The biggest problem with ETAAS has historically been that of interferences. Again, truly spectral interferences are rare. However, since the absorbance of the light by the analytes takes place in a tube, residual matrix components from the ash/char stage may form smoke which obscures the light beam. For some sample matrices, this is not a problem, *e.g.* tap water and river waters that have very little matrix. This is true for some acid digested samples too, but not all. For most sample types, it is therefore necessary to remove as much of the matrix as possible during the ash/char stage. For some analytes that are quite refractory, this is not a problem. Temperatures of 1000 °C may be used with no chance of analyte loss. However, the majority of analytes have a melting point (or even a boiling point) below that temperature and so there is a chance that they will be volatilised away and lost before the atomize stage. This may be exacerbated in sample matrices that have numerous other ions present, some of which may combine with the analyte forming an even more volatile compound. There are numerous possible things to do that can ameliorate this sort of problem. Consulting the manufacturer's handbook/cookbook would be a good starting point. These usually give an indication of the best matrix modifier to use for a certain element and often recommend ash and atomize temperatures. Although these temperatures can deviate slightly depending on the sample matrix, it is a good reference point. Other methods available are given in the useful hints and tips section, below.

4.3.2 Useful Hints and Tips

If a standard operating procedure (SOP) exists for the determination of an analyte in a specific sample matrix, then follow it. Do not spend ages re-inventing the wheel. If an SOP does not exist, then it is worthwhile looking elsewhere before embarking on a significant method development. For instance, many instrument manufacturers produce numerous technical notes that give useful pointers. Similarly, many papers have been published in the literature. It is unlikely that you will be the first person ever to undertake this sort of analysis. However, some methods are trade secrets and, in such cases, information may not be found easily. If this is the case, then there may be no alternative but to do some experimental research and develop a method.

Consultation of the instrument manufacturer's handbook will indicate which matrix modifier to use. A matrix modifier is actually a bit of a misnomer. It is usually a chemical added that stabilizes the analyte to higher temperatures during the ash phase enabling more of the matrix to be evaporated away. Many matrix modifiers exist. Some analytes may require a palladium-based modifier. Here, the palladium forms an intermetallic compound with the analyte stabilising it to higher temperature. Some analytes require magnesium nitrate as a modifier. Here, the magnesium nitrate disintegrates at high temperatures forming magnesium oxide, trapping the analyte within its crystal structure and preventing its loss. Some analytes prefer a mixture of palladium and magnesium nitrate. Numerous other matrix modifiers have been used and reported in the literature, but the palladium and/or magnesium are amongst the most commonly used. The modifier may be added simultaneously with the sample, *i.e.* the autosampler collects 25 µL of sample and then 10–25 µL of matrix modifier before dispensing them both into the graphite tube. An example of a genuine matrix modifier may be used for the analysis of seawater. The presence of huge concentrations of chloride causes dense clouds of white smoke during the atomization stage. The addition of ammonium nitrate as a matrix modifier helps volatilise the chloride away during the ash stage leaving the analytes in the graphite tube and hence facilitating their interference-free determination (or at least reducing it to a level where the background correction system of the instrument can cope).

The first thing to do when developing a method is to ensure that the sample is drying in a controlled and reproducible way. This may be achieved using a mirror and looking through the tube during the dry stages. If a gradual evaporation is occurring, then that is ideal. If the sample droplet is not evaporating, an insufficiently high drying temperature is being used. Increase this by perhaps 20 °C and try again. Repeat this process until the drying is under control. If, however, the sample froths and spits whilst drying, then too high a temperature is being used. Decrease the temperature by 20 °C and try again. It is important that spitting does not occur during the drying stage because there is a chance of analyte loss and, since the tube may not be the same temperature throughout its length, any droplets landing away from the injection point will probably experience a different temperature program. This will lead to poor signal shapes, poor precision and possible inaccuracy. If the sample appears to be drying nicely, but has not dried completely before the ash stage starts, then extend the drying time by 10 s. This may sound a bit "hit and miss", but in reality, optimising the drying temperature does not take long. Often, a two-stage drying is undertaken, with the second 20–30 °C higher than the first. Remember that an acid digest will probably have a higher boiling point than water-based samples. Nitric acid has a boiling point about 20 °C higher than water and therefore a drying temperature of 130–140 °C will be necessary. If organic solvents are to be introduced, the drying temperature may be lower (*e.g.* for pentane, hexane, methanol) or higher (*e.g.* xylene, decalin or white spirit). If organic solvents are used, it is worth considering if they are miscible with aqueous-based matrix modifiers. If they are not, the organic sample should be dried and then the matrix modifier placed on top of the sample residue and dried before the rest of the temperature program continues. Under such circumstances, numerous drying stages may be required. It is also worth checking that an organic solvent does not dissolve or damage the polystyrene autosampler tubes. Do this before placing 30 prepared samples in them!

Once the drying process has been optimized, the ash/char temperature should be optimized. This is achieved by performing an "ash curve". This is where a relatively low temperature is chosen *e.g.* 400 °C, a standard introduced, dried, ashed and then atomized. The absorbance signal obtained during the atomize phase is then plotted on the *Y*-axis against the ash temperature on the *X*-axis. The process is repeated, each time increasing the ash temperature by 100 °C. If no analyte is lost, the signal remains fairly stable. However, at some point, analyte will begin to be volatilised away during the ash stage, leading to a decrease in signal during the atomize stage. It is at this point where the downturn in signal occurs that indicates the ash temperature is too high. The temperature to use should be 50–100 °C lower than this. The presence of the matrix modifier may well extend the useable ash temperature by several hundred degrees. Once this point has been identified using a standard, it is critical that the same process be used for a sample. It may be that at lower temperatures, the smoke produced by un-volatilized sample matrix makes taking an absorbance reading too difficult. However, this problem will lessen as the temperature gets higher until the large majority of the matrix has been removed. Hopefully, the matrix modifier used will extend the usable ash temperature to above the point where the interfering species are removed completely. There is no guarantee of this, but if the matrix modifier can stabilize the analyte to a point where the majority of the interferences are removed, the instrument's background correction system may be able to cope with the remaining interference. Another pseudo-matrix modifier is the addition of air during the ash stage. This can have the effect of burning the organic matrix away. It is especially useful for organic materials, *e.g.* blood and blood products, but will have, at best, a marginal effect on inorganic matrices.

Once the ash/char temperature has been optimized, the atomization temperature should also be optimized. The analyst could just use maximum power and atomize all analytes at 2800–3000 °C. However, this is inefficient and will prematurely damage the tube. What is required is the minimum temperature needed to give a full signal. "Atomize curves" can be constructed starting with a temperature about 100 °C higher than the optimal ash temperature and then plotting the signal (*Y*-axis) against the atomization temperature (*X*-axis). The process is repeated, each time increasing the atomization temperature by 100 °C. The graph plotted should increase steadily and then plateau. The desired atomization temperature is 50–100 °C higher than the point where the plateau starts.

It should be noted that if several analytes are to be determined using a simultaneous instrument, compromise conditions will be required. The ash temperature will have to be set so that the most volatile analyte is not lost and the atomization temperature is sufficiently high for the most refractory analyte to be atomized. It is also worth noting that the analytes to be determined using a simultaneous instrument should require the same matrix modifier. It is usual therefore for analytes to be split into groups, with those such as As, Sb and Se being determined simultaneously and then refractory analytes (if required), *e.g.* Si, Ti, V, *etc.* being determined simultaneously, but in a different analytical run.

These ash/atomize curves have been a staple of ETAAS method developments for many years. It is always worthwhile to perform a spike/recovery test on the sample to identify if any residual matrix interferences exist. Here, a calibration curve is constructed and a sample is analysed using the optimal temperature program. Once the concentration of the analyte of interest has been calculated, the sample may be spiked with a concentration equal to what is already present. This spiked sample is analysed and the concentration is calculated. Ideally, the concentration determined should have doubled (a recovery of 100%). In general, most analysts will accept recovery values of 90–110%. If, however, the recovery is only 70%, then there is still a significant matrix interference effect. If this is the case, then the analyst may be forced into using standard additions as a calibration method. For standard additions calibrations, four points are usually used: a zero addition (unspiked sample), a sample plus a spike of half the determined concentration, a sample plus a spike equivalent to the determined concentration and a sample plus a spike of twice the determined concentration. This should be plotted and the line should be parallel to a normal calibration curve. If it is not, it is confirmation that interferences exist. Unfortunately, if this has to be repeated for all samples, then the workload increases four-fold (*i.e.* each sample must be analysed four different times (plus three replicates of each)). This is far from ideal and many laboratories will try to avoid standard additions, unless there really is no other option. It should be noted though, that standard addition calibration does not correct for all types of interference and so there is still no guarantee of accuracy.

As discussed previously, air oxidation during the ash stage can help remove organic matrices as they are effectively burned off as carbon dioxide. It may also have a small stabilization effect on some analytes as they are transformed into less volatile oxides. If air has been used, it is an extremely good idea to have a second ash period perhaps at a lower temperature afterwards. This is to ensure that all air has been removed from the system before the atomization stage occurs because if air is present at very high temperatures, extreme damage to the graphite tube occurs.

The best way of method validation is through the analysis of a certified reference material. As with all methods and techniques, the CRM should be prepared in the same way as the samples and analysed during the same run and under the same conditions as the samples. If the results are in good agreement with certified values, then it may be concluded that the analysis is accurate and fit for purpose.

Before starting an analysis, it is worth checking the state of the graphite tube. Does it look the same as a new "un-used" one? If so, it is worth continuing with the analysis. However, if there is a clear indication of oxidation, with roughening of the graphite around the injection port, possibly a slightly larger injection port, *etc.*, then this is indicative that the tube is starting to wear and may not be suitable for the analysis to be undertaken. If the tube appears to be fine, use a mirror to check the inside of the graphite tube. Sometimes the pyrolytic carbon can start to flake and if a flake obscures the light beam, the tube will impede the analysis. If the analysis is for Cd in five or six river water samples (plus the calibration), then a tube with clear signs of oxidation may last that long – but it may not. The analyst will have to make a judgement call. If the analysis is for Ti in 30 samples, then a new tube is definitely required. If a tube does disintegrate during an analysis, it is extremely inconvenient. Graphite particles will coat the quartz windows at the ends of the graphite tube housing and these will need to be cleaned. If an optical feedback system is employed to judge the temperature of the graphite tubes, then this will also need to be cleaned. Similarly, the contact electrodes that pass the electricity through the tube will need cleaning. All of these components need periodic cleaning anyway, but ideally not in the middle of an analysis. Once everything has been cleaned and the debris from the ex-tube removed, a new tube can be fitted. The auto-sampler must then be checked that it is aligned correctly, a new calibration run, *etc.*

Many analytes form refractory carbides, Si, Ti, V, W, *etc.* These often inhibit atomization. When such analytes are to be determined a graphite tube in almost pristine condition is required to prevent their formation. Any indication of disruption of the pyrolytic carbon is likely to lead to a decrease in sensitivity that gets progressively worse throughout the analysis as the surface wears further. This leads to excessive signal drift, poor precision, possible carry over from one sample to the next, *etc.*

Aligning the autosampler can be a tricky business. The tip should go through the injection port in the tube making sure there is no contact, however small, with the sides of the hole. If there is contact, even the slightest of brushes against the tube wall, then it is only a matter of time before the autosampler tip gets snagged and bent out of shape. If this is the case, then the autosampler may continue dispensing samples, possibly on the outside of the tube until a QC check discovers a problem. The tip should not go too far inside in case it hits either a platform (if present) or the wall at the bottom of the tube. Either may damage the tube or the tip of the autosampler. Similarly, it should go sufficiently inside the tube so that when the sample and matrix modifier are dispensed, they are not done from such a height that the droplet lands and splashes. A mirror is required so that the analyst can see clearly the positioning of the autosampler tip when in the tube. If only a very limited sample volume is available, *e.g.* 1 mL, then the autosampler can be adjusted so that it goes further into the sample cup to ensure that the sample is collected. Time and effort can be spared if it is ensured from the outset that sample rather than air is being collected. This should be checked prior to analysis. If the depth adjustment was required during a previous analysis and now the cups contain 2 mL of sample, the autosampler should be re-adjusted so that only the end 1 or 2 mm of the tip enters the sample.

Different instruments will have different designs of graphite tubes, different background correction systems, *etc.* The best and most efficient background correction system is that which uses the Zeeman effect. The theory of this is beyond the scope of this text. However, it is necessary to know that it is more efficient at overcoming interferences than conventional deuterium background correction. The one drawback it has is that there is also a reduction in analytical sensitivity.

Modern instrumentation exhibits significantly better performance than those from 30 years ago. The design of the tube is one of the major advances. Tubes may be heated longitudinally (heated from the ends), where there is a temperature gradient with the centre of the tube being the hottest. These are generally more prone to interferences. This is because analyte atoms atomize from the hot centre of the tube into the light beam, condensing at the cooler ends onto the graphite surface where they may be re-atomized into the light beam. This leads to double peaks. Some manufacturers therefore use transversely heated tubes. These are heated from the side and exhibit a much more uniform temperature throughout the length of the tube. Most manufacturers provide the option of buying tubes with an integral platform. These offer considerable advantages over non-platform tubes, be they longitudinally or transversely heated. The use of a platform ensures that the analytes become atomized *via* the hot atmosphere within the tube rather than the heated tube directly. This helps overcome numerous interference types including the production of double peaks and some matrix-based ones. Inevitably, the more complex the tube, the more expensive it is. It is possible to prepare an in-house style platform furnace by breaking up an old tube and finding a fragment that fits loosely into the bottom of a second tube. Care has to be taken to ensure that it does not obscure the light beam but, because of the loose contact with the graphite tube, it does offer similar advantages to those of a commercially available platform tube at much reduced cost, albeit with a comparatively less performance efficiency.

The material the tube is prepared from can also be important. Some tubes are made of electrographite and others from pyrolytic graphite.

The former is less dense and more porous and would provide very poor data for some refractory analytes, *e.g.* those that form stable carbides. The pyrolytic graphite is significantly less porous and dense and provides a surface that analytes cannot easily soak into or react with. Similarly, the pyrolytic graphite also ensures minimal interaction with the sample matrix and is therefore more robust when aggressive or reactive samples are to be analysed.

In general, it pays to use as many facets of the stabilised temperature platform furnace (STPF) concept as possible. This was devised by Slavin *et al.* in the 1980s to ensure maximal reduction in interferences.[3] These are as follows:

- use a good quality pyrolytic graphite tube
- use a platform
- use rapid furnace heating during atomization
- stop the internal flow of argon during atomization (to prevent dilution and rapid removal of analyte atoms)
- use an appropriate matrix modifier
- ensure the instrument has rapid electronic processing (not so much of a problem nowadays as all instruments have this)
- use peak area measurements rather than peak height
- use background correction.

Although the paper by Slavin *et al.* is very old, the STPF concept has been the backbone of successful ETAAS analyses since.

4.4 Data Analysis

As with all analytical techniques, data analysis should be treated as a fundamentally important aspect rather than as an afterthought. There is no point in working up a method, analysing numerous samples and then submitting a report without careful scrutiny of the data that have been obtained. Several aspects need to be examined including the accuracy as well as inter- and intra-sample precision. Accuracy can be measured in several different ways, the best of which is the analysis of a certified reference material that has a matrix as closely matched to the sample being analysed as possible. Analysis of such a material will test the efficiency of the digestion/extraction/preconcentration procedure as well as the accuracy of the actual analysis process. It would also test the correctness of the mathematics used to calculate the concentration in the original sample, *i.e.* taking into account all dilutions or pre-concentrations that have occurred. It should be emphasised that the material should be as closely matrix matched as possible. If leaves are the sample of interest, then there is a very big difference between tomato leaves that have a "furry" texture and holly leaves that have a waxy cuticle. The efficiency of an acid extraction/ digestion could vary significantly between the different types of leaf material. Similarly, with soil samples, the soil could be chalky, peat-like, sandy, *etc.* Fortunately, certified materials covering many types of material are available. If a material identical to that being analysed is not available, then a similar one would certainly be better than nothing. At least one certified material should be prepared and analysed per batch of samples. The certificate should give a mean concentration of each analyte along with their associated uncertainty values. Ideally, the concentrations determined experimentally (once any moisture content has also been taken into account) should lie within the uncertainty boundaries of the CRM. If not, then something may have gone wrong in one or more steps. If the concentration is exactly a tenth of the expected concentration, then that usually signifies a calculation error or omission of a dilution factor during the calculation. Similarly, a two-fold error could signify something as silly as 25 mL volumetric flasks that are usually used not being available and so 50 mL ones were used instead. This can often occur if a new experiment on the instrumental computer is made from a previous one. It is easy to overlook the column headed "sample volume". The analyst should therefore look to see if two, five or ten-fold errors have occurred, since these are the most common during calculations. If the result of the CRM is accurate for some analytes but not for others, it would indicate that the actual analysis is fine, but that the sample preparation procedure is not ideal for all analytes. This may therefore need modifying to obtain more accurate data. An alternative possibility is that the standards were made correctly for some analytes but not others. The use of independent check standards should pick that error up during the analysis though.

Some laboratories insert the volume the digested sample has been diluted to, any further dilutions and the mass used into the instrumental software and allow that to do the calculations. Care should obviously be taken to ensure that no typographical errors are introduced as these can be difficult to identify. For instance, if an analyst enters a weight of 0.3526 g instead of the intended 0.2526 g, a significant error of about 40% would clearly result. This would possibly be easier to identify than had 0.2626 g been inserted instead of 0.2536 g. The latter would result in an error of ~4%. Although not a huge error, it could still confound results. The values inserted to the software should therefore be checked carefully prior to the analysis. If an obvious error is identified after the analysis, the software will enable the correct values to be inserted and the results re-calculated, but this may be inconvenient if the instrument is being used by another analyst.

An alternative method to determine the accuracy would be to use a completely independent technique, *i.e.* to prepare the sample differently and then analyse them using a different instrumental technique. This approach is usually not adopted because it is very time-consuming and, for some laboratories, an alternative analytical technique may not be available. A method of accuracy assessment that is often adopted is that of spike/recovery experiments. This is where a sample is spiked with a known concentration of analytes before the acid digestion occurs (if a solid sample is to be analysed) and simply mixed to ensure homogeneity if it is a liquid sample. An un-spiked aliquot of sample is prepared simultaneously to the spiked. If the un-spiked concentration was 4 mg kg^{-1} and a spike of 5 mg kg^{-1} was added, then the spiked sample should give a total of 9 mg kg^{-1} because 100% of the spiked amount was recovered. Good accuracy is assumed if the recovery is 90–110%. In this example, if the spiked sample gave a total of 8 mg kg^{-1}, then only 40 of the 50 mg kg^{-1} spike has been recovered, giving a recovery value of 80%. The analysts should then rely on the standard operating procedure for that laboratory to see what to do next. Some laboratories may do a 100/80% correction for that analyte, whereas others will leave the data as produced and state that only an 80% recovery was obtained. It should be noted that the concentration of the spike should be similar to that which is there naturally. If a concentration of 5 mg kg^{-1} is present naturally, then a spike of 2–8 mg kg^{-1} would be appropriate. Spike concentrations of 0.1 or 50 mg kg^{-1} would not be suitable for a natural content of 5 mg kg^{-1}.

Some laboratories may analyse in-house materials to assess the accuracy of the analysis. This is usually a bulk material that has been analysed against a CRM and values assigned for each analyte, *i.e.* if the correct answers for the CRM were obtained, it is assumed that they are also correct for this in-house material. These materials are particularly useful when the use of a CRM is limited, *e.g.* if the supplier allows the purchase of only a few bottles per laboratory per year, or if they cost so much that their routine use is prohibitively expensive. The in-house material may then be used for extended periods at minimal cost assuming it does not degrade with time. The results for the in-house material may then be plotted on Shewhart-style quality control graphs.

Intra-sample precision may be measured using three (or more) replicate analyses of the same sample digest. Clearly, the same sample should give a very similar signal for all replicates. If it doesn't then there is a possibility that the sample had not reached the flame AAS instrument before measurement commenced. The first replicate would therefore be significantly lower than the others. Ideally, this should be spotted during the analysis, so that it may be stopped, the uptake time modified to include the first replicate and then re-started. If the error was not spotted during the analysis, then it may be possible to omit the first replicate and take an average concentration from the remaining ones. Here, it would be an advantage if four or five replicate measurements were made. If a replicate is significantly higher than the subsequent ones, it may indicate a partial blockage of the nebuliser or flame slot has occurred. This will continue (and possibly get worse) until the next check standard (which should be run every 10 samples) is run. If this is too low, the obstruction can be removed, the instrument re-calibrated and the previous 10 samples re-run. An alternative scenario is that the sample is consumed completely before all the measurements are obtained. Here the signal/concentration decreases with increasing number of replicates. Under such circumstances, the check standard may well be within 10% of its nominal value and other samples may have good intra-sample precision. Errant replicates can be removed from the calculations.

Inter-sample precision may be measured by comparing the average concentration of the analytes in different digests or aliquots of the same sample. If these differ significantly, it may well indicate that the sample preparation procedure is not wholly under control. If this is the case, then either the procedure being used could be optimised further to try and identify which part is causing the trouble, or a different procedure may be adopted.

References

1. *An Introduction to Analytical Atomic Spectrometry*, ed. L. Ebdon, E. H. Evans, A. Fisher, and S. J. Hill, John Wiley & Sons, Chichester, 1998, ISBN-10:047197417X.
2. I. N. Pasias, N. I. Rousis, A. K. Psoma and N. S. Thomaidis, *At. Spectrosc.*, 2021, **42**, 310–327.
3. W. Slavin, D. C. Manning and G. R. Carnrick, *At. Spectrosc.*, 1981, **2**, 137.

5 Inductively Coupled Plasma: Optical Emission Spectrometry

5.1 Introduction

Inductively coupled plasma–optical emission spectrometry (ICP–OES) also known as ICP–atomic emission spectrometry (ICP–AES) has been around since its development in the 1960s. However, the technique really started to become routine in the 1970s. Numerous instrumental developments have occurred during that time and so now ICP–OES is regarded as being one of the industry standard and most reliable techniques. A historical perspective, some of the theory behind it, some instrumentation and some applications may be found in numerous books, for example, one edited by Hill.[1] In common with the other chapters of this book, this chapter is not going to go into the theory in any great detail. Instead, it too will concentrate on the advantages and disadvantages of the technique, describe many of the common faults and errors encountered and provide some useful experimental hints and tips. However, a very brief overview may be useful to the reader. A sample is usually introduced as a liquid *via* a nebuliser and spray chamber assembly. This transforms the liquid into an aerosol where the smallest droplets are transported by argon gas to the ICP. The ICP is a partially ionized gas that is at very high temperature. This dries, atomizes and can ionize the analytes present in the aerosol in a similar manner to the flame of a flame AAS

Practical and Technical Guides for Laboratory-based Chemists No. 1
Atomic Spectrometric Methods of Analysis
By Andrew Fisher
© Andrew Fisher 2025
Published by the Royal Society of Chemistry, www.rsc.org

instrument. The atomized/ionized analytes are thermally excited at temperatures of between 5000 and 10 000 °C and so electrons in the analytes are temporarily promoted to more energetic levels. As they lose their energy and return to the lower energy states, light with wavelengths characteristic of that element is emitted. The intensity of the light at these characteristic wavelengths is directly proportional to the concentration of that analyte.

Historically, a huge range of spectrometers that were of different designs, *e.g.* Czerny–Turner, Ebert, Paschen–Runge, *etc.*, with either sequential detection using a photomultiplier or simultaneous detection using several photomultipliers were used. The sequential instruments obviously took longer to do an analysis for several analytes and would consume more sample. Historically, the simultaneous instruments would have a number of photomultipliers arranged on a Rowland circle at a position at which light from a specific analyte would be diffracted. Since these were fixed into position, the laboratory would have to ask the manufacturer to set it up for them and then if an analyte that was not amongst those the instrument was set up for requires determination, big problems could occur. More recently, most manufacturers have now settled on instruments that contain an Echelle-type spectrometer with either a charge coupled device (CCD) or charge injection device (CID) detector system. Again, how these work is beyond the scope of this chapter, but may be found elsewhere.[2] However, it can be said that both detector systems offer simultaneous detection of numerous analytes and, along with the spectrometer, are compact and have quite high spectral resolution. The reference given above is very elderly and the technology has improved significantly. The basic physics of how they work has not changed though, so the reference is a good place to start for anybody wanting to learn about them.

The measurement of the emission from the plasma may be achieved in two ways: radial (where the plasma is usually upright and measurements are taken from the side) and axial (where the plasma is usually horizontal and measurements are taken from the end). Some dual view instruments may use both – but not at the same time. Briefly, the axial orientation has better sensitivity and provides lower limits of detection (by a factor of 3–5, depending on the element). However, it is more prone to self-absorption effects at very high concentrations than a radial instrument, *i.e.* it has a shorter linear range regarding concentration compared with the radial viewed plasma. It often has the small disadvantage of requiring an extra gas flow to shear the plasma to prevent damage to the detector, to prevent the re-formation of molecules in the cooler region of the plasma and to help minimise self-absorption.

5.2 Advantages and Disadvantages of ICP–OES

The advantages and disadvantages of ICP–OES instruments are summarised in Table 5.1 and will be discussed in greater length in the text.

5.2.1 Advantages

Instruments of either axial or radial configuration have a long linear range that spans typically five orders of magnitude. The limits of detection vary depending on the analyte, whether or not the analyte's most sensitive wavelength is used and whether or not it is an axial or radial instrument. For a radial instrument, limits of detection can range from 0.1 $\mu g\,L^{-1}$ for the most sensitive analytes, *e.g.* Mg and other alkaline earth elements up to 20–50 $\mu g\,L^{-1}$ for elements such as As, P, S and Se. Single figure $\mu g\,L^{-1}$ values are obtainable for most analytes. It is certainly true to say that ICP–OES offers significant sensitivity improvements over flame AAS for refractory elements such

Table 5.1 The advantages and disadvantages of ICP–OES instruments.

Advantages	Disadvantages
Usually simultaneous detection of numerous analytes/analyte lines	Quite expensive to purchase and operate.
Low detection limits (typically $\mu g\,L^{-1}$ range)	Requires specialist facilities, *e.g.* fume extraction
Long linear calibration curves	Some limitations to good/suitable detection limits for some analytes/ sample types
May be used to analyse samples in solid, liquid or gaseous state	May be prone to EIE effects, dependent upon their concomitant concentration
Most have a long spectral range (180–850 nm), enabling most analytes to be detected (>60).	Some samples produce spectra that are very line rich, leading to interferences dependent upon the resolution of the spectrometer.
Consumes a low volume of sample.	Often need expensive service contracts
Offers the possibility of measuring line emission from ions as well as atoms	May need specialist training courses for novice users
Easily automated	
Reduced chemical (matrix) interferences compared with a number of alternative atomic spectrometric techniques (XRF, ETAAS, *etc.*)	

as Al, Si, Ti, W and Zr. These analytes form very stable oxides (amongst other molecules) and are not readily broken down into the atomic state at temperatures used in flame AAS (~2800 °C for a nitrous oxide–acetylene flame and only 2000–2300 °C for air–acetylene). However, the 'temperature' of the inductively coupled plasma is 6000–8000 °C and hence atomization/ionization is significantly improved. As an example, the limit of detection (LOD) for Al using flame AAS is approximately 1 mg L^{-1}, whereas it is single figure µg L^{-1} for ICP–OES (when using the most sensitive wavelength). Axial instruments typically have LOD between three and five times lower than radial ones. The plasma is so energetic that it enhances atomization compared with flame AAS and also manages to ionize many analytes. This means that there is a much larger choice of wavelengths available to use analytically. By convention, atom lines are given the notation (I), *e.g.* Mg(I) at 285.2 nm and ion lines are given the notation (II), *e.g.* Mg(II) at 279.80 nm. The line rich spectra can lead to spectral interferences. This will be discussed later.

Most instruments are of a 'simultaneous-detection' design nowadays. This means that numerous analytes or several wavelengths of the same analyte may be determined in the same time it takes to measure only one. Therefore, up to 60 analytes may be determined in about a minute. Other, slightly older instrument types are either rapid sequential or pseudo-simultaneous, but can still measure perhaps 10 analytes in less than a minute.

The multi-element capability has several advantages. It enables a much smaller volume of sample to be consumed. This is because the uptake rate for a typical ICP–OES nebuliser is approximately 1 mL min^{-1} meaning that a suite of analytes can be determined in volumes as little as 1–2 mL. It also means that if an analyte is to be determined in a sample and the analyst has no idea what concentration it is present as, then multiple wavelengths of the analyte can be monitored that cover very many orders of magnitude. For instance, for Fe, the most sensitive wavelength could be used which could cover the concentration range of 5 µg L^{-1} to perhaps 10 mg L^{-1} and a less sensitive wavelength could also be used that covered the range of 1–500 mg L^{-1}. If required, a further wavelength could be chosen that went to even higher concentrations. In this way, the linear range can be extended to perhaps seven or eight orders of magnitude. This is facilitated by the instrumental software that enables different wavelengths to be programmed with different concentrations of standard. Sometimes, if an analyst thinks that a certain wavelength may potentially suffer from interferences, then several different wavelengths of the same analyte that have a similar sensitivity may be monitored. A spectral interference on one wavelength is less likely to cause problems for the other wavelengths. Hence, if the answer obtained for the three different wavelengths is the same, then the analyst can be fairly confident that no spectral interferences exist. If the result from one wavelength differs markedly from the other two, then the data from this wavelength can be ignored since interference effects would appear to be present.

The long linear range is a big advantage over the absorbance-based techniques. If only one wavelength of an analyte has been used, then if the concentrations found in a set of samples are above the top standard and assuming that the top standard is not at the top of the linear range, then another standard of higher concentration may be prepared, measured and added to the calibration curve. In this way, numerous time-consuming dilutions can be avoided. One note of warning though: if an error has been made during the preparation of the calibration standards, e.g. a 10 000 mg L^{-1} stock solution was used instead of 1000 mg L^{-1}, the error is less easy to spot. Under such circumstances, in an absorbance-based technique, all of the calibration standards are likely to be above the linear range and hence a very curved or even flat calibration would result. An order of magnitude error on a calibration that can extend to five orders of magnitude, e.g. on an ICP–OES instrument, may still be on the linear part of the curve. Only the results of a quality control standard or certified reference material may indicate that something has gone awry.

The large majority of analyses undertaken using ICP–OES are of liquid samples or solid samples that have been brought into the liquid state by an acid dissolution/extraction. Here, the sample is typically transported from the sample container via a peristaltic pump to a nebuliser and spray chamber assembly where it is transformed into an aerosol (a suspension of the liquid sample as fine droplets in the argon carrier gas) and then transported in a flow of gas to the plasma. However, both solid and gaseous samples may also be analysed. Solid samples may be analysed directly in one of two ways: electrothermal vaporization and laser ablation. Both of these need an accessory to enable the introduction. For the former, a few mg of sample is weighed onto a platform which is then heated electrothermally in a similar fashion to electrothermal AAS (and would suffer the same problems associated with it). A programme of temperatures is then applied in which as much of the matrix is removed as possible prior to the atomization of the analyte(s) to the plasma in a flow of argon. While it does enable a pseudo-matrix removal, it is a slow process, often leads to poor precision because of the difficulty in weighing out exactly the same mass of sample, carry-over effects, *etc.* Although used in some research laboratories, there are few routine laboratories that use this methodology. For sample introduction using laser ablation, a very expensive laser ablation unit is required (often a similar price to the basic ICP–OES instrument itself). Modern ablation units can focus the laser onto extremely small areas (often no more than a few microns in diameter). They can therefore be extremely useful for analysing microscopic inclusions in alloys or in rocks. It can also be regarded as being a minimally damaging method of analysis because the craters it leaves are almost invisible to the naked eye. This would, of course, also be an advantage for sample types that are precious, *e.g.* archaeological ceramics and glasses, artworks, forensic samples, *etc.*, or for single cell analysis of biological materials. As well as the drawback of expense, there is always a problem with being able to calibrate adequately since it requires a calibration standard to be as perfectly matched in matrix components to the sample as possible. In addition, the analysis of such small areas can lead to problems with being able to sample representatively. Problems associated with repeatability and reproducibility may also arise. The use of these techniques is beyond the scope of this text, but mention has been made so that the reader is informed of their capabilities. The use of vapour generation as a means of sample introduction is discussed at length in Chapter 8. However, it is useful at this juncture to state that it may readily be coupled with ICP–OES instrumentation for the determination of analytes in numerous sample types. During this method, the analytes are transported to the plasma as a gas, hence leading to increased sensitivity and some separation from matrix components.

Most instruments are capable of detecting analytes over the spectral range of 167–852 nm (Al to Cs). This means that pretty much all metals, metalloids and many non-metals are capable of being detected. It should be noted that for those elements that have a very low wavelength, *e.g.* Al at 167 nm and As, P, S and Se, all of which have primary wavelengths below 200 nm, a flow of argon (or a vacuum) should be used to remove the air from within the spectrometer. This prevents absorption of the emitted light by molecules such as carbon dioxide, nitrogen and other molecules, hence improving their sensitivity significantly.

5.2.2 Disadvantages

A typical ICP–OES instrument is roughly in the middle range of expense for atomic spectrometric instrumentation. However, the cost of running it can be quite high. This is because it consumes typically 13–15 L min^{-1} of argon, meaning that a full-size gas cylinder will last 6–8 hours. Liquid argon can be purchased but requires an extremely large Dewar flask for storage and hence needs ventilation. A drawback of liquid argon is that even though Dewar flasks are quite good at insulating the gas, there is still some boil-off. If the Dewar is filled just before a holiday, then when operators return a week later, it could be half empty meaning that a severe cost has been incurred for no gain.

The laboratory facilities should also be taken into account. ICP–OES instruments require fume extraction to remove argon, all plasma by-products, any ozone produced and heat. As previously noted, argon gas also needs to be provided. This can be through standing cylinders with pipework directly to the instrument, or cylinders or a Dewar of liquid argon kept in a separate store with the gas piped up through the building to the instruments. Although more expensive to install, the latter option is usually preferred because cylinders tend to be quite heavy and have considerable surface contamination, hence transporting them around the building can be problematic. The pipework used should also be taken into account. Nylon or other plastic piping can be used but can allow some other gases to permeate through them. Stainless steel is therefore often the material of choice. The use of single cylinders of argon can be problematic. They may last 6–8 hours when an instrument is in use, but if a series of shorter experiments are undertaken, then there is a possibility that the gas may run out halfway through an experiment. This can potentially damage the torch or will at least be annoying because the experiment will have to be stopped, the instrument turned off and the gas cylinder changed. The instrument would then have to be turned on again, stabilised to ensure all air has been removed, re-calibrated, *etc.* Therefore, many placed now use manifolds in which ten cylinders are arranged in two banks of five. If one bank runs out of gas, an automatic switch-over to the next occurs. Checking the state of the gas cylinders every few days prevents errors and enables an empty bank of five cylinders to be replaced.

Although detection limits tend to be better than those of flame AAS, they are still too high for the determination of analytes in some sample types. The analysis of seawater, for instance, is problematic for two reasons. The very high concentration of dissolved salts is likely to cause blockage of the injector tube of the torch after prolonged use, and secondly, other than the major ions present, *e.g.* Na, K, Mg, Ca and S, other elements tend to be present at extremely low concentrations, often <1 μg L^{-1}. Therefore, if more sensitive instrumentation is not available, lengthy preconcentration procedures are required prior to their determination. Similarly, if rare earth elements are to be determined in environmental samples, the heavier elements, *e.g.* Lu are usually well below the detection limit once the material has been dissolved.

As stated previously, the very energetic plasma is very good at atomizing and ionizing elements. This can cause problems when easily ionized elements (EIE) are present at high concentrations in the sample. Although this is not going to be discussed at great length, a brief explanation is required. If K is to be determined in a sample that contains a significant amount of Na, then the equilibria below are set up:

$$K \leftrightarrow K^+ + e^-$$

$$Na \leftrightarrow Na^+ + e^-$$

If there is a large amount of Na present in the sample (or some other EIE), there becomes an excess of electrons in the plasma and so the equilibrium is pushed towards K rather than K^+. Since K and K^+ emit light at different wavelengths, this can cause a problem if the calibration standards do not also contain high concentrations of the concomitant EIEs. This is because an enhanced signal for K would be obtained. It may therefore be necessary to add a suitably high concentration (*e.g.* 200–500 mg L^{-1}) of an easily ionized element, *e.g.* Cs, to both the standard and sample solutions to prevent this from happening. Care should be taken not to add too much of the EIE so as to cause salting up of the nebuliser *etc.*

5.3   Common Errors and Problems

5.3.1   Calibration Errors

Since calibration standards are required for quantitative analysis in most ICP–OES applications, the section at the beginning of the AAS chapter which discusses the preparation of standards, is equally applicable. There is one other note on the linear range of these instruments that should be clarified. Although the linear range can extend to perhaps five orders of magnitude, if samples of extremely different concentrations are analysed during the same experiment, errors will result. For instance, attempting to analyse a sample that contains 0.2 mg L^{-1} Cu when the previous sample contained 400 mg L^{-1}, will result in a serious carry-over effect, even with the fastest clearing nebuliser and spray chamber assemblies. Similarly, if a series of samples have a similar concentration of an analyte, then an appropriate series of standards should be made. For instance, if the analyte is present at 5–10 mg L^{-1} in the samples, then there is little point in making a calibration curve with standards of 0, 20, 40, 80 and 100 mg L^{-1}. Instead, standards of perhaps 0, 1, 2, 4 and 10 mg L^{-1} would be much more appropriate. This can make standard preparation slightly trickier since different analytes will have a very different concentration in acid extracts of soils or sediments. For instance, Fe could be present at many hundred mg L^{-1} whereas Cu or Zn may be present at ~1 mg L^{-1}. Preparation of a mixture of analytes all at a range up to 500 mg L^{-1} would lead to gross errors for those analytes present at lower concentrations. This is because those standards of very high concentration will have a huge lever effect at the lower concentrations and so any error, even one of 5%, could skew the bottom of the calibration significantly. An analyst should therefore think very carefully before using a stock standard that has a mixture of analytes all at the same concentration.

Once the analyst has prepared the standards, they can think about using the instrument. A schematic diagram of an ICP–OES instrument is shown in Figure 5.1. This is to aid the reader since much of the following text refers to individual components of it. The figure shows a radial-style plasma, with measurements taken from the side. An axial instrument has the same components but takes readings from the end of the plasma. Ideally, the instrument should be turned on to "warm up" at least 20–30 minutes before use. This will give the

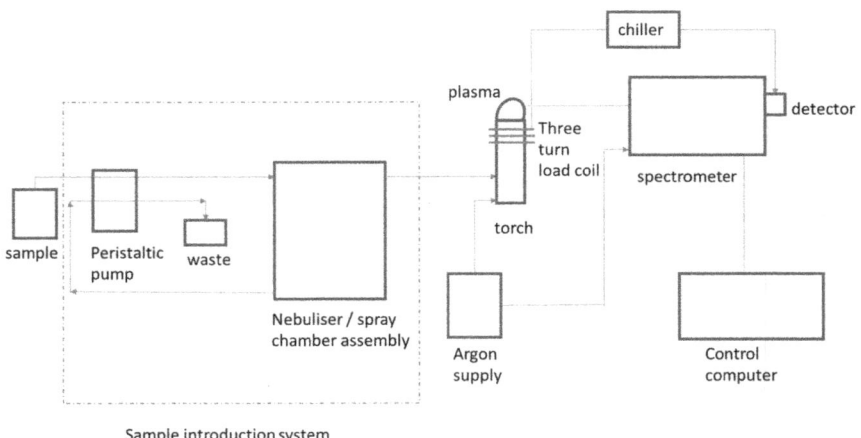

Figure 5.1 A schematic diagram of an ICP–OES instrument.

detector time to be chilled to the correct temperature and the electronics to stabilise. If an instrument is turned on and then an attempt is made to use it immediately, a drift in sensitivity often results. It is therefore a good idea to turn the instrument on before making the standards since this can save time later. However, it is important to note that if a trickle of argon has not been left passing through the spectrometer overnight, a small build-up of moisture can occur on the detector for some instrument types. For those instruments that require the detector to be chilled, this should be removed by turning on the argon flow for 15–20 minutes before attempting to ignite the plasma. Failure to remove this moisture before igniting the plasma could lead to the detector (which may need to be chilled to well below 0 °C) having a layer of ice formed on it. This can potentially cause significant damage and require the detector to be replaced. Depending on the manufacturer and instrument model, this can cost thousands of pounds and will require a visit from a service engineer.

As stated previously, some elements (Al, As, P, S and Se) have extremely low primary wavelengths and require a flow of argon to pass through the spectrometer to improve sensitivity by decreasing the amount of light absorption by molecules in the air. This can take a while even if the flow is hastened. A way to overcome this problem is to keep a constant trickle of argon through the spectrometer, even when the instrument is not in use. Although this obviously increases the consumption of argon, it can save a significant amount of time. If the determination of these analytes is not frequent, then with forward planning, the analyst can schedule for the analysis the next day and just leave a trickle of argon going overnight. Unfortunately, the usual 15–20 minutes of argon flow before the plasma is ignited, although adequate for moisture removal, will not be sufficient to remove all traces of the air and if an analysis is undertaken before all traces of the air have been removed, then significant drifts in sensitivity would result.

5.3.2 Sample Introduction Errors

Many laboratories have a standard operating procedure for turning the instrument on. This should always be adhered to. Failure to do so can lead to very silly (and embarrassing) errors. For instance, an analyst will regret reporting to the laboratory manager that the ICP "isn't giving any signal" only to discover that the peristaltic pump tubing used to transport the sample to the sample introduction system has not been clipped into place on the pump correctly or if it is so old it has collapsed or split. An unwary novice analyst may even plumb the tubing incorrectly and pump the sample directly to waste or accidentally pump the waste bottle contents to the nebuliser. This can be problematic for a number of reasons; not least that it wastes time, but it may also use up samples that have very limited volume. It is wise to therefore check that the sample is pumping to where it is supposed to be. In any of these cases, the sample would not be transferred to the instrument. The pump tubing should always be inspected prior to use to ensure that it shows no significant signs of wear and tear. A previous user may have left it clamped in place overnight (or possibly over a weekend). This can also cause damage to the tubing.

It is standard practise for the pump tubing that transports the sample from the nebuliser/spray chamber assembly to waste to have a slightly wider bore than that used to transport it to the nebuliser. This is to ensure that the spray chamber does not "flood", *i.e.* start to fill with liquid. If the spray chamber does flood with a mixture of samples and wash solution, then reliable results will not be produced because as flooding progresses (before it gets to the carrier tube at the base of the torch), the mass transport efficiency will change resulting in an increasing drift effect between calibration and sample sensitivity. If this goes unnoticed, then the flooding will increase until the liquid enters the torch and the plasma is extinguished. This will mean having to dismantle the nebuliser/spray chamber assembly, drying the torch out, re-assembling it all and then re-starting the analysis. Many, but not all, instruments have an interlock that monitors the peristaltic pump tubing in the waste channel for bubbles and, if none are observed, it stops the pump.

The performance of the pump tubing may be assessed by deliberately admitting an air bubble to the sample input and the waste exit tubing and then watching its passage to ensure it moves smoothly and that there are no pauses or that it does not gently rock backwards and forwards. If there are pauses, the screw on that channel of the peristaltic pump can be tightened or loosened until an improvement in flow is observed. If no improvement is seen, then a fresh piece of tubing could be used. The screw on that channel of the pump should be loosened until there is no flow and then tightened very slightly until it starts again. In this way, any deterioration of the pump tubing can be assessed and minimised.

Junior analysts should remember that, in general, if an instrument was working perfectly well the previous time it was used and then does not when they try to use it, then more times than not, it is a minor error that can be rectified relatively easily. The reliability of instruments has improved markedly over the last 10 years and so the chances of a serious random fault developing overnight have very much diminished.

5.3.3 Physical Errors

If the torch has been removed for cleaning, then it should be dried thoroughly before it is replaced in the instrument. This can be achieved using tissue paper or by using a rapidly evaporating solvent, *e.g.* acetone, ensuring adequate ventilation and safety precautions are in place. Remnants of this should also be allowed to evaporate before replacing the torch. If a wet torch is installed, then problems will be experienced during plasma ignition. Instead of a stable plasma forming, an extremely noisy and unstable discharge occurs. In an extreme case, it is possible that this can damage the torch, sometimes irreversibly.

The plasma is formed through a spark from a Tesla coil partially ionizing the argon gas. These ions then get caught in the magnetic field of a three-turn load coil, colliding with other argon atoms and ionizing some of them. The result is a cascade effect that produces a self-sustaining plasma. The load coil is water cooled and usually made of copper. Many nowadays have a polymer sheath. With age, the sheath can become worn and a new coil may need to be fitted. If a very wet torch is inserted and the plasma attempted to be lit, then the load coil as well as the torch can potentially be damaged.

5.3.4 Assessing the Performance of the Instrument

Once the instrument has been turned on and allowed to warm up, many laboratories undertake a performance report. This protocol is pre-programmed into most modern instruments and usually involves the measurement of a mixed standard of analytes. The performance report usually checks that a minimum signal for each of the elements is obtained. It then checks that precision is adequate and the limits of detection are sufficiently low. If one or more elements fail on any of these three parameters, then the analyst can decide whether to continue with the analysis or to try and rectify the problem. For instance, if the minimum Cu signal required is 15 000 counts per second, but only 14 200 counts per second are achieved, then the analyst could say (with some justification), that no serious problem exists and that a minor drop in sensitivity is insignificant when a full calibration is run. If it is only one element in the performance check solution that has failed and it is not one of the analytes of interest to be measured during the analysis, an analyst could take the approach of "not a problem". If, however, only 220 counts per second were obtained instead of 15 000, then there is clearly a problem. In such a case, it is likely to be an error in the setting up of the instrument (usually somewhere in the sample introduction system) and other elements will equally be adversely affected. Alternatively, the solution used for the performance check could have been prepared incorrectly. Manufacturers do supply these solutions, but some laboratories prepare their own to save extra costs. If the majority of the analytes have one or more failures then, if that laboratory does not have any rules in place, *e.g.* a standard operating procedure, for such a circumstance, it will be a judgement call by the analyst as to what course of action to take. Their judgement should be guided by how badly the elements have failed, how many of them are the analytes of interest, *etc.* The elements monitored in the performance check often cover the entire spectral range and have different sensitivities.

5.3.5 Wavelength Selection Errors

The analytes of interest are then chosen from the periodic table. On selecting an analyte, a large number of potential wavelengths are usually provided within the instrument's selection programme, normally along with their sensitivity and potential interferences. As with all analyses, it helps enormously if the analyst already knows what is in the sample and so can judge the potential interferences. For instance, if an analyst knows that there is a huge amount of iron present in the sample, they would do well to try and avoid choosing an analyte wavelength that has an iron wavelength close by – even if it is one of lesser sensitivity. For many industrial samples where the sample is a product of manufacture, this is usually not a problem because often they are checking that the product meets manufacturing specifications and so is already fairly well characterised. Similarly, most biological samples are relatively well characterised, for instance, it is well known that blood contains significant amounts of Fe, Na, K, Ca *etc.* However, it is also known that huge concentrations of rare earth elements and other elements that have a large number of emission wavelengths are not expected. Environmental and geological samples can be more problematic. In general, for soils, sediments and rocks, Al and Si can be relied on to be at extremely high concentrations, but depending on the nature of the sample, Ca, Fe and Mg may or may not also be present at percentage levels. For example, an "ordinary" soil may contain 2–5% Fe $(20\,000–50\,000 \text{ mg kg}^{-1})$. However, a very sandy or chalky soil or a peat may have significantly less. Assorted ores will obviously have extremely high concentrations of certain metals too. The unpredictability of environmental samples can therefore be problematic when estimating potential interferences.

Most modern instruments have good spectral resolution and can separate peaks to 5–20 picometres, However, if a concomitant (or even an analyte) is present at exceptionally high concentration in the digested sample, then tailing from its signal peak may impinge on the signal from another analyte. It is therefore wise, if possible, to choose an analyte wavelength that is at least 0.1 nm from potential interferences. There are occasions when this is not possible and so then either a matrix separation protocol may be undertaken prior to analysis or manipulation of the signals in the form of judicious selection of background correction settings may be made. An example of this is shown below. The signal for the analyte (Cu at 324.754 nm) in a standard is shown in Figure 5.2a. The shaded regions on either side of the analyte peak are the left and right background correction points, with the area above these points being taken as the signal for the analyte. Figure 5.2b shows the Cu signal at 324.754 nm for an acid extraction of a sediment sample. Here, there is clearly a concomitant that is interfering with the left-hand correction point leading to a significant shift in the background correction. On this occasion, it is a very insensitive Fe wavelength at 324.716 nm that is causing the problem. Figure 5.2c shows how changing the background correction points can affect the result. By moving it slightly away from the concomitant peak, a much more accurate signal evaluation results. If unchecked, this would have led to an underestimate of the true Cu concentration of approximately 13%. Far worse scenarios can be experienced, but this one is sufficient to demonstrate that the spectra around the analytes should be checked for the samples as well as for the standards.

5.3.6　Errors from Organic Solvents

If organic solvents are to be analysed, *e.g.* dilutions of oils or oil fractions, organic extracts of materials, *etc.*, then it is worth noting that standard peristaltic pump tubing may dissolve or start to disintegrate in some solvents. If this does occur, then the solvents can enter the working parts of the peristaltic pump and cause significant damage. Tubing especially designed for organic solvents is available and should be used if

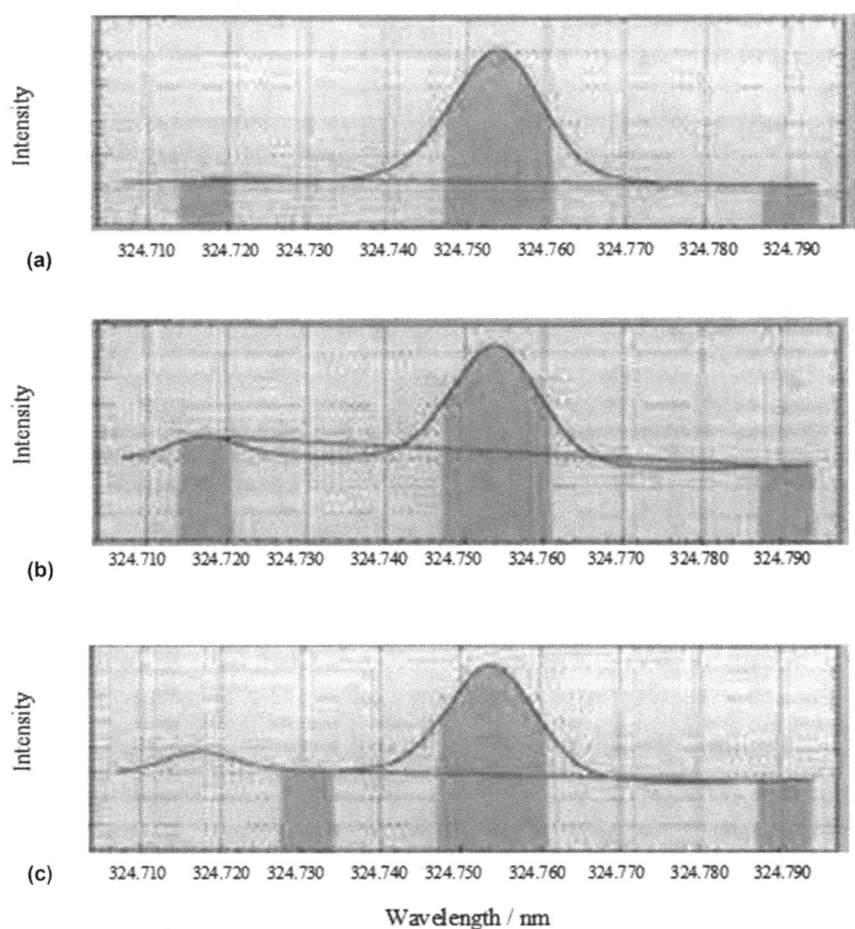

(a)   324.710   324.720   324.730   324.740   324.750   324.760   324.770   324.780   324.790

(b)   324.710   324.720   324.730   324.740   324.750   324.760   324.770   324.780   324.790

(c)   324.710   324.720   324.730   324.740   324.750   324.760   324.770   324.780   324.790

Wavelength / nm

Figure 5.2   (a)-(c) The effects of background correction on the signal of an analyte.

disintegration of the normal tubing is a possibility. Historically, plasmas are not at all keen on organic solvents being aspirated into them. Some would just be extinguished as the instrument's plasma load matching network could not react fast enough, others would expend much of their energy in breaking down the solvent rather than atomizing and exciting the analytes. Fortunately, with the huge advances made with modern instrumentation, these problems are less significant than they were. In addition, specialist organic solvent introduction accessories which involve spray chambers that are chilled to well below 0 °C are available. These decrease the vapour pressure of the solvents and hence decrease the organic loading of the plasma.

If organic solvents are to be analysed, it is necessary to matrix match the standards, *i.e.* the standards should also be prepared in the same solvent. Although some solvents are miscible with water, *e.g.* methanol, ethanol, acetic acid, *etc.*, others are not. For those that are miscible with water, simple inorganic stock standards may be employed. However, some care needs to be taken regarding the solubility/stability of the dissolved components when doing so. Adding water to organic solvents and *vice versa* can affect solubility However, for those solvents that are not miscible with water, *e.g.* hexane, dichloromethane, *etc.*, specialist organic analyte stock solutions must be used. Organic samples should never be analysed against standards prepared in inorganic solvents because the nebulisation efficiency and plasma dynamics are completely different. Although this can be overcome, to some extent, using an internal standard, it is far safer to matrix match.

Matrix matching is not confined to organic solvents. Any reagent used to extract a material, *e.g.* enzymes in a buffer solution, reagents used during the BCR three stage extraction procedure for soils and sediments, alkaline extraction media, *etc.*, will have a different viscosity, surface tension and hence nebulisation efficiency when compared with standards prepared in dilute acid. Again, the use of internal standardisation can help compensate for these differences, but matrix matching samples and standards is always the best approach – especially when used in conjunction with an internal standard. The use of internal standards will be discussed later in the text.

Many laboratories operate under ISO accreditation schemes (*e.g.* ISO 9001:2015 or ISO 17025). As identification of risk is the focus of these schemes, then common errors and problems such as those discussed above may be addressed by using standard operating procedures and protocols within those frameworks. Training both in-house from senior analysts and from instrument manufacturers will also help a junior analyst to understand the instrumentation better which, in turn, should prevent many of the errors.

5.4 Hints and Tips

As with all instrumental techniques, there is a mass of information in the scientific literature that could be consulted before an analyst spends a great deal of time and energy working up a method only to find that there are standard methods of analysis or it has all been done before. Useful annual updates on new techniques or instrumentation may be found in the Atomic Spectrometry Update by Evans *et al.*[3] Other applications may be found in other updates of the same series including for biological, food and clinical samples,[4] environmental samples[5] and industrial-based samples, *e.g.* glasses, polymers, ceramics, nano-materials, electronic components, ferrous and non-ferrous samples, organic materials, *etc.*[6] In all of these updates, it is not just ICP–OES applications that are discussed, with ICP–MS, LIBS and XRF applications also described. However, it is a good place to start for any analysts new to the area. In addition to the literature, the instrument manufacturers will have an applications department where useful advice may be sought. Many of them also publish application notes that may assist a new (or even experienced) user.

5.4.1 Operating Conditions

The operating conditions employed are usually a compromise that works for all analytes but are optimal for none. If a suite of analytes is to be determined and one is suspected to be present at very low concentrations, it is a good idea to optimise the operating conditions for that element so that it maximises the chances of determining it successfully. The key parameters to optimise tend to be the nebuliser gas flow rate, the viewing height (for radial instruments and sampling depth for axial ones), and the power applied to form the plasma. The optimisations may be undertaken using a univariate or multivariate approach, with the latter being preferable because the parameters are inter-dependent.

Similarly, when analysing organic samples, the operating conditions can differ significantly from those used for aqueous-based materials. Along with the organic sample introduction accessory, some manufacturers supply a semi-demountable torch whose injector is of a different bore to that used for aqueous samples. It should be noted that normal "inorganic" operating conditions are likely to be adequate for supplying data, but they will be far from optimal and hence decreased sensitivity and poorer limits of detection would result. This is of most importance when determining analytes at low concentrations. The accessories that specialise in the introduction of organic solvents usually operate at a significantly reduced temperature, *e.g.* -5 °C, so that the vapour pressure of the solvent is diminished. This helps to stabilise the plasma and increases sensitivity. As discussed previously, it should be noted that the peristaltic pump tubing used should be solvent resistant rather than the standard material.

5.4.2 Choosing the Correct Nebuliser/Spray Chamber Assembly

A plethora of different nebulisers and spray chambers have been developed over the years. Some nebulisers are designed to introduce samples with high dissolved solid loading. Others are better at introducing particulates, *e.g.* in slurries. The introduction of slurries or particulates can be especially difficult for some nebuliser types as they are easily blocked. Even unfiltered waters can prove to be problematic. A blocked nebuliser may lead to low pressure unions, *e.g.* between peristaltic pump tubing and PTFE transfer tubing to become detached. If the analyst does not notice this, sample(s) may drip somewhere and be lost. This is clearly a nuisance because it will need cleaning up and waste time but is even more problematic if limited sample was available in the first instance. Although accessories now exist for un-blocking nebulisers (effectively, a syringe that can be filled with ethanol that is used to back-flush the nebuliser), it is still annoying to have to remove the nebuliser mid-analysis, clean it, flush the ethanol traces from it, re-introduce it to the spray chamber and then re-establishing the argon gas flow. This is doubly annoying not just because of the interruption to the analysis, but also if the nebuliser is removed and not returned to exactly the same position in the spray chamber, different sensitivity may be obtained. This may necessitate a re-calibration. It is also possible to extinguish the plasma if the nebuliser gas flow is turned on too quickly because a burst of air (or possibly ethanol if it has not been removed completely) will reach the plasma. This will obviously require a new ignition sequence. If an autosampler is in use, it is possible that a blockage may not be noticed for some time, *e.g.* until an alarm sounds because a periodic check sample is out of the acceptable range, which is also annoying. The choice of nebuliser to use is therefore dependent on the application. The ones that are most easily blocked tend to be self-aspirating, *i.e.* they do not require samples to be pumped to them. They therefore have the advantage of having no noise associated with the peristaltic pumps. The pump noise originates from the rollers of the pump rolling over the peristaltic tubing causing a squirt of liquid sample to pass through. It can be minimised by having a larger number of rollers and for the pump to be going faster. A spray chamber also helps dampen the noise, with bigger volume spray chambers dampening the noise more efficiently but at the expense of requiring longer washout times between samples. However, the self-aspirating nebulisers still produce a more stable signal and require only a small spray chamber to separate larger droplets from smaller ones. The analysis can therefore be much quicker when using the self-aspirating nebulisers because the sample need only travel 30 cm through tubing to the nebuliser rather than pass an extra 30 cm of pump tubing and a length of transfer tubing. Washout is also more rapid because of the small spray chamber being used and because there is less tubing to wash. However, it should be noted when using self-aspiration that liquid should be aspirated for as much time as possible and the time where air is allowed to enter should be kept to a minimum. If a lengthy period of air is introduced, the plasma may be extinguished.

The function of spray chambers is two-fold: to separate large sample droplets (sent to waste) from the smaller ones (transported to the plasma) and to dampen pump noise. Spray chambers also come in assorted designs, with the traditional Scott double pass ones now out of favour with the faster cleaning cyclone style chambers being used more frequently. However, a spray chamber with a larger volume helps dampen the pump noise more than a smaller one. Spray chambers should be incubated to ensure a stable temperature is maintained. This may be achieved using a cooling water jacket or, more recently, a Peltier-cooled holder. If the temperature of the spray chamber is not kept constant, then sensitivity will drift badly as more sample will be vaporized at elevated temperatures. This can be more problematic for laboratories that are not temperature controlled and where the temperature can change from 25 to approximately 40 °C over the course of the day.

It should be noted that nebuliser and spray chamber designs and the materials from which they are made can differ depending upon the sample type being introduced. It would be unwise to use a glass nebuliser and spray chamber assembly (and a quartz torch) if hydrofluoric acid remnants are in the sample digests. Therefore, if geological samples that have been digested in hydrofluoric acid are to be analysed, then either the hydrofluoric acid must be boiled off on a hotplate in a specialist fume cupboard, or the fluoride component removed using a boron-containing material (boils off as BF_3). Failing either of those, a polymer nebuliser/spray chamber assembly and a ceramic-based torch could be used. Other sample types also de-vitrify the glassware significantly. These may include samples containing strong bases, *e.g.* sodium hydroxide. Again, polymer-based introduction systems may prevent damage.

Overall, the sample introduction system is an extremely important set of components and should not be ignored for successful analyses.

A good recent tutorial by Nelms and Kutscher discussed the components of the sample introduction systems of both ICP–OES and ICP–MS instruments.[7] Most workers use a system where the sample needs to be pumped into it. This is partly because many laboratories use an autosampler. These are excellent when they work, offering complete automation and enabling analysts to perform other tasks while they undertake the analysis. As with the AAS autosamplers, they can be programmed to run check standards interspersed with the samples and then the computer control can be programmed what to do if they are not within the acceptable range, *etc.*

5.4.3   Calibration Standards

Most instruments now have the flexibility to enable extra calibration standards to be added to the sample list if required, *e.g.* if some of the samples have concentrations above the top standard. Similarly, if the analyst has had a slip of the finger and has incorrectly programmed the concentrations of the standards, most instruments will simply allow this to be corrected – even after the analysis is complete. The only thing they cannot help the analyst with is if they have inadvertently forgotten to include an analyte in the element list to be determined. In such a case, a new analysis would be required.

5.4.4   Internal Standards

The following paragraphs discuss the use of acids as %, *e.g.* 2% nitric acid. Although concentrated nitric acid is 68% v/v, 2% nitric acid normally refers to 2% of the concentrated acid, *i.e.* 2 mL of concentrated acid diluted to 100 mL rather than ~3 mL

The use of an internal standard can improve long-term stability significantly. This is normally an element or elements that is/are not in the sample at an appreciable concentration and so can be added to all standards, samples, blanks and quality control samples. The analyte signals can then be normalised to the internal standard(s) and so any drift in sensitivity can be corrected for – assuming the drift occurs for the analytes to the same extent as the internal standards. Drifts in sensitivity may arise for many reasons, *e.g.* the samples were removed from a fridge and not allowed to warm to room temperature prior to analysis. As the samples slowly warm up, sensitivity may increase as the samples' viscosity decreases. Another scenario is that an acid-digested sample will have a very different viscosity and surface tension to standards prepared in 2% nitric acid because it is likely to have 5–10% acid and the dissolved material. Transport efficiency through the nebuliser/spray chamber may therefore be very different, hence causing a change is sensitivity rather than a long-term drift. Again, the software is usually flexible enough to choose which internal standard to use for which analyte either before, during or after the analysis. A mixture of internal standards being used for some elements is usually also an option. Internal standard(s) may be added using a pipette to all samples and standards or, they may be introduced *via* a T-piece and then mixed with the samples before the nebuliser. The latter option is more time-efficient but assumes that sufficient mixing occurs. The manual add-ition to all samples using a pipette requires the volume of the sample to be known so that the same concentration of analyte can be added to each. This is not always the case and so errors can be made by an unwary analyst. A way around this problem would be to pipette a known volume of every sample into tubes so that the volume of the internal standard required can be kept constant. It is then necessary to mix the samples (by inversion or vortex-mixing) to ensure complete homogen-ization of the sample and internal standard.

In general, many manufacturers suggest that keeping the dissolved solid level below 0.1% m/v in samples aids the stability of the analysis. This is undoubtedly true and is the reason that samples prepared by fusion and some acid digestions often have to undergo a further di-lution prior to analysis. In the case of fusion sample preparation, 0.25 g of sample may be mixed with 1 g of flux, fused and then dis-solved in 100 mL of dilute acid. This would lead to an overall dis-solved solids loading of just over 1% m/v. A 10-fold dilution is therefore required otherwise the injector of the ICP torch will slowly become blocked leading to signal drift. Eventually, complete blockage occurs, necessitating instrumental shutdown and cleaning. For acid extractions of soils and sediments, 0.25 g of material can be extracted in 5 mL of nitric acid or aqua regia and, once extraction is complete, filtered into 25 mL volumetric flasks. This is effectively a 100-fold dilution, but since the material does not dissolve completely, the dissolved solids loading is much less than 1%. A small drop in sen-sitivity results, and this is partially due to the high acid level but may readily be corrected for by the use of an internal standard(s). However, the dissolution of steel samples (*e.g.* 1 g) in 30% sulfuric acid (20 mL) followed by dilution to 100 mL will have a 1% solids loading and will require further dilution – especially if there are numerous samples for analysis. This would decrease the dissolved solids loading as well as the acid concentration.

5.4.5 Wash Solutions

In general, most laboratories use a dilute acid (*e.g.* 2% nitric acid) to wash the sample introduction system between samples. This can be modified for certain analytes that adhere to glass quite strongly. For instance, when determining Hg, nitric acid takes a very long time to wash it away. However, thiourea or other sulfur-containing reagents are far more effective.[8] Although this reference is elderly and discusses ICP–MS detection, the sample introduction systems are similar to those used for ICP–OES and the reagents are equally applicable.

To prevent a build-up of contaminants in the wash solution, some analysts wipe the uptake tubing with some tissue immediately after the analysis of a sample is completed and then immerse the tubing in the wash solution. Care should be exercised to ensure that no leaching of analytes occurs from the tissue when samples that have a strong acid or alkali content are being analysed. Other analysts dip the uptake tube in a separate container of water or 2% acid prior to the wash solution. Either way is often acceptable and preserves the useable life of the wash solution.

5.5 Quality Control

As with all analyses, a standard should be analysed every 10 samples. This will give the analyst an indication whether there is any instrumental drift and will give them the opportunity to correct for it, to recalibrate, to ignore it, *etc.* Most analysts will continue as long as these check standards are within 10% (or ideally 5%) of their initial value. This is an important point to note: it should be their initial value that the checks are compared with rather than their nominal value. Consider a standard of nominal value 10 mg L^{-1}, if there was a slight error during preparation, then its experimental value may be 11 mg L^{-1}. However, when included with the other standards, a decent calibration is obtained. If this particular standard is then run as the check and a value of 9 mg L^{-1} is obtained, then it is within 10% of the nominal value, but significantly more than that (18%) from the experimental value. The analyst should not be lulled into a false sense of security by making such a basic error. This 18% difference could be very significant and the results obtained could potentially be significantly higher than calculated.

The best way of ensuring the overall analysis (from sample preparation to reporting) is accurate and reliable would be to analyse a certified material. As always, this should have been prepared and analysed in exactly the same way as the samples and be as closely matrix-matched to them as possible. Unfortunately, certified materials for all sample types do not exist and so compromises sometimes have to be made. If a sandy soil is to be analysed, then ideally a sandy soil certified material should be analysed. If one is not available, then any soil (or even a sediment) certified material would be better than nothing. Clearly, a blood or leaf-based material would be inappropriate.

The use of ICP–OES has been a staple of many laboratories for several decades and is likely to stay that way for many years to come. It is possible that as the sensitivity and calibration accuracy of LIBS instruments increases, that ICP–OES instruments may slowly become less popular.

5.6 Data Analysis

In general, the same procedures as those adopted for the analysis of AAS data (Section 4.4 of this book) should be adopted. Certified materials should be analysed and run at the same time as the samples and the data obtained from them compared with certified values. In the absence of suitably matrix matched certified materials, then spike/recovery experiments could be used to assess accuracy. In common with the analysis of AAS data, it will be up to the individual laboratory's own good practice whether or not any analytes that are lower than certified values are corrected by the recovery factor.

The use of an internal standard can help minimise errors causing poor precision. It should be noted though that if a partial blockage of the nebuliser has occurred, which is much more likely than in flame AAS, the internal standard may not be reliable. Correction using an internal standard helps correct for any long-term drift and is efficient even if the signal sensitivity has drifted by up to 20–25%. However, a drop of 90% in the signal would indicate that the internal standard is multiplying the analyte concentrations by a factor of 10. Any slight inaccuracies will also therefore be multiplied by 10. This means that if the instrument raw counts indicate that there is 2.5 mg L^{-1} instead of 2.7 mg L^{-1} then once the internal standard has been taken into account, it will read 25 mg L^{-1} instead of 27 mg L^{-1}. This may not seem much of a difference. However, it should be emphasised that the problem gets worse the closer to the LOD the measurements are.

If the peristaltic pump tubing starts to fail mid-analysis, then transport of the sample to the plasma may take longer (or cease completely). Checking the replicate measurements is therefore necessary to identify problems such as this. A slight failure of the tubing would potentially manifest itself as a much lower first replicate concentration compared with subsequent ones. A complete failure of the pump tubing or a blockage somewhere in the sample introduction system would lead to close to zero raw counts being obtained for all analytes and the internal standard for all subsequent replicates.

It should be noted therefore **that simply looking at the mean concentration can lead to numerous errors**. If three replicate measurements of the same sample are 7, 23.5 and 23.2 $mg\,L^{-1}$ because of a problem with the sample reaching the plasma in time for the first replicate measurement, the mean concentration will be given as 17.9 rather than the more likely 23.35 $mg\,L^{-1}$.

Analysis of a certified reference material shortly after calibration may help identify errors during standard preparation. If, for instance, the experimental value for a CRM is one-tenth of the certified value, it could be that a dilution error has occurred, either through using an incorrect pipette or the stock standard was $10\,000\ mg\,L^{-1}$ rather than $1000\ mg\,L^{-1}$ as had been expected. Alternatively, a simple calculation error may have occurred. The calculations and stock solutions can be checked, re-made as necessary and then re-run on the instrument. It is clearly easier for this to be done during the analysis rather than returning to it some days later.

References

1. *Inductively Coupled Plasma Spectrometry and its Applications*, ed. S. J. Hill, Wiley-Blackwell, 2nd edn, 2006, ISBN 978-1405135948.
2. J. M. Harnly and R. E. Fields, Solid State Array Detectors for Analytical Spectrometry, *Appl. Spectrosc.*, 1997, **51**(9), 334A–351A.
3. E. H. Evans, J. Pisonero, C. M. M. Smith and R. N. Taylor, *J. Anal. At. Spectrom.*, 2022, **37**, 942–965.
4. M. Patriarca, N. Barlow, A. Cross, S. Hill, A. Robson, A. Taylor and J. Tyson, *J. Anal. At. Spectrom.*, 2022, **37**, 410–473.
5. J. R. Bacon, O. T. Butler, W. R. L. Cairns, O. Cavoura, J. M. Cook, C. M. Davidson and R. Mertz-Kraus, *J. Anal. At. Spectrom.*, 2023, **38**, 10–56.
6. S. Carter, R. Clough, A. Fisher, B. Gibson and B. Russell, *J. Anal. At. Spectrom.*, 2022, **37**, 2207–2281.
7. S. Nelms and D. Kutscher, *Spectroscopy*, 2022, **37**, 51.
8. Y. F. Li, C. Y. Chen, B. Li, J. Sun, J. X. Wang, Y. X. Gao, Y. L. Zhao and Z. F. Chai, *J. Anal. At. Spectrom.*, 2005, **21**, 94–96.

6 Inductively Coupled Plasma: Mass Spectrometry

6.1 Introduction

Since its inception in the mid-1970s, inductively coupled plasma–mass spectrometry (ICP–MS) has become an industry standard technique. There are several versions, *e.g.* single quadrupole, triple quadrupole (ICP–MS/MS), high resolution instruments that use magnetic sectors, similar instruments that have multi-collector detectors, *etc.* They all have their advantages, disadvantages and abilities, meaning that they are often suited to different applications. Each of the techniques will be discussed in this chapter. Since many of the faults and errors are common to all types, this chapter will start with the simplest (single quadrupole instruments) and then move on to the more complex ones, discussing their additional problems. In common with other chapters in this book, the theory of each will not be discussed in great detail. However, an extremely good introduction to the techniques is given in a book by Thomas.[1] The book covers the main theoretical aspects of sample introduction, ion formation, ion extraction, ion focussing, collision and reaction cells, ion separation using assorted mass spectrometers and detection in a logical and well-presented way. Also covered were different methods of quantitation, a review of interferences, routine maintenance, and alternative sample introduction techniques and several chapters described different application areas. In addition, a briefer account of some of the

Practical and Technical Guides for Laboratory-based Chemists No. 1
Atomic Spectrometric Methods of Analysis
By Andrew Fisher
© Andrew Fisher 2025
Published by the Royal Society of Chemistry, www.rsc.org

other analytical atomic spectrometric techniques was also provided. Another readable, but older, account of single quadrupole instruments was presented by Linge and Jarvis.[2] A recent article by McCurdy and Yamanaka described the key aspects of setting up a method for ICP–MS/MS.[3]

Although a comprehensive description of the theory is not necessary, a brief overview of how many of the instruments work may be of assistance to the reader and will therefore be provided. Figure 6.1 depicts a schematic diagram of a single quadrupole ICP–MS instrument. The schematic diagram shows an instrument that has a linear configuration. Many modern instruments have a right angled-configuration, where the mass spectrometer and detector are vertically above the ion lens system. This is useful for two reasons: it gives the instrument a smaller footprint enabling more instrumentation to be placed on a bench and it maximises the ion transmission to the detector. Many older instruments used the ion lens system to guide the ions around a photon stop or up over a step. The instrument then has to re-focus the ions towards the mass spectrometer and detector. Ion transmission is more efficient if they are just "turned" 90° and the light-stopping ability is as efficient as the other configurations. Since ion transmission is maximised, the instruments tend to have greater sensitivity and lower limits of detection.

For all ICP–MS instrument types, liquid samples are usually introduced using a conventional nebuliser/spray chamber assembly, similar to those used for ICP–OES. The aerosol produced is then transported to the plasma by a stream of argon. It should be noted that, although most analyses will involve the introduction of a liquid sample, it is possible to analyse solid samples directly. This could be through the use of laser ablation (LA) or, if bulk analysis is required, through electrothermal vaporization. The plasma dries, atomizes and

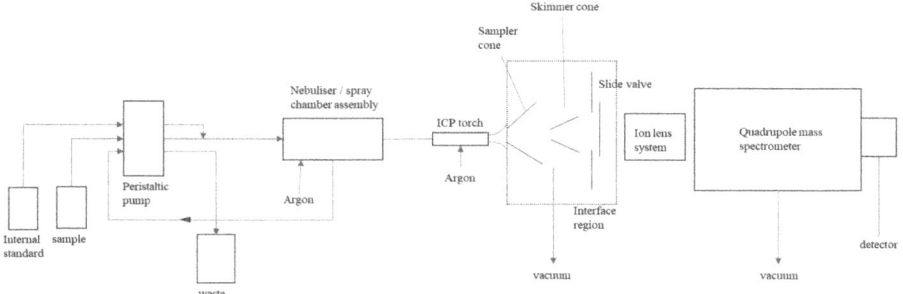

Figure 6.1 A schematic diagram of a quadrupole ICP–MS instrument.

must then ionize the analytes. In ICP–OES, the analyte need not necessarily be ionized. However, for ICP–MS measurements, it is a prerequisite since the detection system works on a mass-to-charge ratio. If there is no charge then there will be no signal. The extent of ionization is one of the determining factors of the overall sensitivity and ranges from 100% ionized for Cs (which has the lowest first ionization energy) through to F, which has a higher ionization potential than the argon gas used to form the plasma. This means that F cannot be determined directly using conventional ICP–MS.

Once ionized in the plasma, the ions pass through a partially evacuated interface region which joins the atmospheric pressure plasma to the high vacuum mass spectrometry system. The interface comprises a "sampler" cone through which the ions extracted from the plasma pass, a partially evacuated chamber and a "skimmer" cone. As the ion beam from the plasma passes through the sampler cone, it cools rapidly and expands adiabatically, *i.e.* it forms a cone of ions. Some of these then pass through the skimmer cone, through a slide valve and are focussed by an ion lens system to the mass spectrometry system. The design of the ion lens system varies between manufacturers but is designed to focus as many ions to the mass spectrometer as possible while ensuring that stray light is prevented from impinging on the detector. The mass spectrometry system separates the numerous ions according to their mass-to-charge ratio (m/z). Many ICP–MS instrument types enable one m/z to be detected at any one instant with the others being passed to waste. The time that each m/z is measured is called the dwell time and is typically ~10 ms. The detector is often an electron multiplier and is capable of detecting all m/z; hence, the necessity to separate them temporally so that only one m/z reaches it at a time. The mass spectrometry system then changes its parameters, stopping that m/z from being detected and allowing another to pass to the detector. The majority of instruments are therefore rapidly sequential in nature, enabling perhaps 70–80 analytes to be measured in less than a second. This is called a "sweep". The whole measurement process is repeated numerous times with many sweeps being obtained before the signals are summed for each analyte and presented as a measurement replicate. The number of sweeps obtained is input to the instrumental software by the analyst. Many do not deviate from the default values set by the manufacturer. Increasing both the dwell time and the number of sweeps can often improve precision if concentrations close to the detection limit are to be determined. This would be at the expense of an increased analysis time. As with most other analytical techniques, it is usual for several (at least three) replicates to be measured per sample.

As well as its ionization potential, another factor that governs the sensitivity of an analyte is the number of isotopes it has. If the overall signal for an element is split over six isotopes, it will be less sensitive than if it had all of its signal at only one isotope, even if five of the isotopes summed to only 20% of the total abundance. It should also be noted that, in general, analytes of higher mass tend to have greater sensitivity than those of low mass. This is important to realise because it can have a significant effect when isotope ratio analyses are being undertaken. This is discussed in somewhat more detail later.

6.2 Advantages and Disadvantages of ICP–MS

The advantages and disadvantages of single quadrupole instruments are given in Table 6.1 and are discussed further below.

6.2.1 Advantages

All types of ICP–MS instruments have a very long linear range; spanning up to nine orders of magnitude. If fitted with a Faraday cup as well as an electron multiplier, it may even reach 11 or 12 orders of magnitude. For single quadrupole instruments detection using a multiplier is split over two modes of detection: analog and pulse counting. To ensure linearity, these two modes need to be cross-calibrated. This will be discussed further in the errors section later. Despite it having such a huge linear range, not even the most optimistic of analysts should expect a sample containing 100 mg L^{-1} to not have a carry-over effect on the next sample that may contain a low concentration, *e.g.* 0.01 µg L^{-1}.

Table 6.1 Advantages and disadvantages of single quadrupole ICP–MS.

Advantages	Disadvantages
Long linear range	High cost of purchase
Very low detection limits	High running costs
Can analyse solid, liquid or gaseous samples	Suffers from several interference types
Quasi-simultaneous determination of >80 analytes	Considerable infrastructure required
Some interference types may be overcome using a collision cell	Use of a collision cell decreases sensitivity – especially for low mass elements
It can yield isotopic information	Less tolerant of sample matrix than ICP–OES
	Poor resolution

The detection limits obtainable with modern instrumentation are lower than virtually any other technique. As discussed earlier, detection limits vary between analytes, but can reach as low as 0.01 $\mu g\,L^{-1}$ for a routine analysis. Lower limits of detection may be obtained under certain circumstances which are discussed below. The main limiting factor nowadays is contamination. This can arise from contamination through the use of impure chemicals or through aerial deposition if a laboratory is dusty. It is therefore necessary to use ultra-pure reagents (or use methods such as sub-boiling distillation to clean reagents up) to obtain the best LOD. Some sample manipulations may be undertaken in laminar flow hoods. These hoods help decrease aerial contamination. Some elements are more abundant than others in the environment and therefore these tend to have greater issues with contamination. Contamination may be decreased further by placing the instruments or undertaking sample manipulations in "clean rooms". The detection limits obtainable may be decreased if preconcentration methods are used or if vapour generation is employed to introduce the analytes. A recent paper by Woods and McCurdy provided five practical tips for minimising contamination enabling the lowest limits of detection to be obtained.[4]

As with ICP–OES, the samples may be introduced as solids, liquids or gases. The large majority are introduced as liquids, meaning that solid samples undergo some sort of a dissolution or extraction process prior to the analysis. Attention to the dissolution techniques should be paid to try and ensure that any acids used do not cause interferences during the analysis stage. As with ICP–OES, the alternative to a dissolution/extraction is the introduction of the sample by laser ablation (if very small areas, *e.g.* inclusions in steels or geological materials are to be analysed) or, if bulk analysis is required, by slurry nebulisation or by electrothermal vaporization. The preparation of slurries was discussed in Chapter 2. It is also possible to introduce gaseous vapours into ICP–MS instruments such as when determining As, Pb, Se, *etc.*, using hydride generation. This sample introduction methodology is discussed at length in Chapter 8.

Another advantage is that well over 80 elements of the periodic table may be determined routinely using ICP–MS. The exceptions are the gaseous elements such as the noble gases and some of the extremely heavy transuranic elements. The determination is not simultaneous for a quadrupole-based instrument. Instead, it is very rapid sequential, with a dwell time of typically 10 ms for each isotope monitored. The dwell time may be altered, with longer dwell times, *e.g.* 80 ms, offering better sensitivity and often, better precision. This would come at the expense of increased time required for the determination. For a suite of 20 analytes and for a dwell time of 10 ms, the analysis would take approximately 200 ms per sweep. The number of sweeps can then be chosen to determine the total length of the analysis.

One of the big advantages of ICP–MS is that it offers the possibility of obtaining isotopic information for samples. For an analyte such as Pb, four isotopes may routinely be determined: *m/z* 204, 206, 207 and 208. For a single quadrupole instrument, the precision is inferior to multi-collector-magnetic sector instruments. These offer the best option with regard to isotopic precision measurements. Single quadrupole instruments can still be useful for identifying isotopic anomalies in samples and for isotope dilution analysis. The theory of isotope dilution analysis is beyond the scope of this text. However, it is regarded as being an extremely accurate method of analysis. It can be achieved by mixing an isotopically pure (or with an accurately known enrichment) aliquot of a standard with a sample. An equilibration time is required to enable the aliquot to mix with the sample naturally so that it mimics the same analytes present in the matrix. An extraction may then be made and even if extraction is not 100% efficient and assuming the analyte in the spike behaves in the same way as it does in the natural material, it enables a very accurate analysis. Single quadrupole instruments do not measure more than one isotope at a time and hence the precision will always be poorer than that obtained when simultaneous measurements are made.

6.2.2 Disadvantages

One of the main disadvantages of the technique is its high cost of purchase. A single quadrupole instrument is significantly more expensive than an ICP–OES instrument, with triple quadrupole instruments being up to 50% higher still. Magnetic sector instruments with multiple collector detectors can be a further 3 or 4 times higher than a triple quadrupole. The costs of the instruments are therefore prohibitively expensive for some smaller laboratories. The running costs are also appreciable because they use a similar amount of argon as ICP–OES instruments and have additional components *e.g.* the sampler and skimmer cones on the interface to purchase. It also requires significant infrastructure for installation. Gas storage areas, fume extraction to remove hot argon and ozone and potentially clean rooms are all required.

The main disadvantage of the technique is the large number of interference types that exist. The main type of interferences is called polyatomic. This is where two or more elements combine to form a molecule with a mass nominally the same as an analyte. Numerous examples exist and were catalogued many years ago by Evans and Giglio[5] and may also be found online.[6] An example of a two-bodied interference is argon chloride ($^{40}Ar^{35}Cl^+$) on As at m/z 75 ($^{75}As^+$). Similarly, examples of a three-and four-bodied polyatomic interferences are: $^{32}S^{16}O_2{}^+$ on $^{64}Zn^+$ and $^{32}S^{16}O_2{}^1H^+$ on $^{65}Cu^+$. The concomitants can arise through the sample matrix or through the acid(s) used to dissolve solid materials. Since H and O arise in the presence of water and N and O may be entrained from the air, the acid least likely to cause trouble is nitric (HNO_3). Other acids, such as hydrochloric acid (HCl) and sulfuric acid (H_2SO_4), can easily combine with each other or with the plasma gas forming numerous polyatomic interferences, $e.g.$ $ArCl^+$ (on As^+), $^{35}Cl^{16}O^+$ (on $^{51}V^+$), $^{37}Cl^{16}O^+$ (on $^{53}Cr^+$), $S_2{}^+$ (on $^{64}Zn^+$), $etc.$

Another type of interference is isobaric. This is where two elements nominally have the same mass, $e.g.$ $^{113}Cd^+$ and $^{113}In^+$. Although neither has a mass of exactly 113, the poor resolution of a quadrupole instrument (it has unit mass resolution) will not be able to separate them. Most analytes have more than one isotope, and these isobaric interferences are well characterised and can usually be circumvented. A third type of interference is that of doubly charged ions. The extent of this depends on an analyte's (or a concomitant's) second ionization energy. These interferences are easily understood and avoided, with one of the major culprits being $^{137}Ba^{2+}$, which could potentially interfere with $^{69}Ga^+$ and $^{68}Zn^+$. A recent publication discussing the amelioration of doubly charged interferences may be found at this website.[7]

A fourth type of interference, that is often overlooked, is that of tailing of analyte signals into adjacent analytes. An example would be for the determination of $^{55}Mn^+$ in the presence of very high concentrations of Fe, such as that found in blood samples. The signal from the $^{56}Fe^+$ and, to a lesser extent $^{54}Fe^+$, could merge into the $^{55}Mn^+$ signal artificially raising it. Since Mn is mono-isotopic, there is not even the possibility of using a non-interfered isotope. The instrumental software should inform the analyst of all common potential interferences on the isotopes of each analyte. This should help the junior analyst. As always, it helps enormously if the analyst knows what is in the sample to start with so that they can identify the potential interferences and choose the analyte isotopes accordingly.

Often this is not problematic because the composition of industrial-type samples, *e.g.* glasses, polymers, ceramics, *etc.*, that have been made in a factory's production line will be well known to the analyst working in that factory's laboratory. Natural samples such as rocks, ores, soils and sediments tend to be less predictable.

There are also physical interferences, similar to those experienced in ICP–OES analyses. These are related to differences in viscosity of matrix composition between samples and standards *etc.* and were discussed in the ICP–OES chapter (Chapter 5). It should be noted that most interferences, especially the most problematic type of poly-atomic, may be overcome using modern instrumentation and the collision/reaction cells found in them. The one drawback of using a collision cell is that it decreases the sensitivity. The extent of the re-duction is dependent on sample mass, with the heavier analytes being affected significantly less than light ones. For example, an optimised signal of a 1 μg L^{-1} Pb may give 200 000 counts per second (cps) without a cell, but if a collision cell gas is used, this may decrease to 180 000 cps. However, for a much lighter element, *e.g.* Li, the count rate may decrease from 20 000 to about 50 cps. Fortunately, it is possible to use the collision cell for some analytes and not for others during the same analytical run, but this comes at the expense of re-quiring extra time. This is because it takes a while for the collision cell to be evacuated after it has been used before the next set of analytes that do not require it are determined.

One final problem that ICP–MS analyses suffer from is that the instruments are significantly less tolerant of complex sample matrices than ICP–OES. The nebuliser and plasma torch may become blocked (as with ICP–OES), but the interface region may also become clogged with salts or soot if an organic solvent is introduced. If heavy deposits do occur in this region, then the instrument needs to be shut down and then cleaned. This is not convenient during an analysis. Worse still, if the orifice of the sampler cone slowly becomes clogged, then signal drift can be appreciable. An internal standard may help counteract this, but eventually, the loss of signal will become so se-vere that an internal standard cannot be relied upon. Extremely complex matrices, *e.g.* seawaters or industrial brines, can cause im-mense problems with ICP–MS analyses. Therefore, instrumentation has been developed that enables their introduction. The so-called "aerosol dilution" methodology was discussed recently by Duyck *et al.* in a tutorial review.[8] This useful review, containing 109 references, discusses how the aerosol between the spray chamber and the torch is diluted with gas, provides the theory of the process and gives numerous applications covering several sample types. Using such methodology, dilution of samples with dilute acids is often not necessary and decreases sample manipulation (and hence both time and the possibility of contamination).

6.3 Problems and Errors

Many of the errors associated with ICP–MS are the same as those experienced for ICP–OES but it also has numerous others. Certainly, those of sample introduction to (and removal from) the instrument, *i.e.* incorrectly plumbed peristaltic pump tubing, worn or defective tubing, blocked nebulisers, *etc.* are equally applicable. In addition, the problem associated with placing the sample uptake tube directly into the wash solution and the resulting contamination is also in common with ICP–OES. A further similarity is the trouble of carry-over from one sample to the next. This occurs when a sample containing a low concentration of an analyte is analysed immediately after a sample that had a relatively high concentration of the same analyte. Fast cleaning nebulisers and cyclone-style spray chambers can increase the speed of wash-out. However, only an extremely optimistic analyst would hope that washout would be complete in 30 s for a sample four orders of magnitude lower in concentration than the previous one. The reader is directed to the relevant section of the ICP–OES chapter to understand more about these problems. Similarly, quantitation using ICP–MS normally requires the preparation of calibration standards. Therefore, the discussion of errors that may occur during the preparation of standards, calibration of balances and pipettes, *etc.*, at the beginning of the atomic absorption chapter (Chapter 4) is also applicable to ICP–MS analyses. Choice of an appropriate range of standards is also a problem in common with ICP–OES. A lazy analyst may prepare a series of calibration standards of a mixture of analytes, all with the same concentration range. However, any skewing of the calibration through error of the highest concentration standard will lead to serious errors for analytes present at much lower concentrations. As an example, when a suite of 12 analytes is to be determined and standards of 0, 5, 20, 50 and 100 $\mu g\,L^{-1}$ are prepared for all analytes because the expected concentrations of many analytes fall mainly into this range. However, if the expected Cd concentrations happen to be 0.1–0.3 $\mu g\,L^{-1}$, the concentrations of the calibration standards are inappropriate. Any slight error in the 100 $\mu g\,L^{-1}$ standard would skew the bottom of the calibration curve significantly, possibly to the extent where negative concentrations may be obtained. Therefore, standards of different concentration ranges should be prepared for different analytes. In this example, standards for Cd should perhaps be 0, 0.05, 0.1, 0.2 and 0.5 $\mu g\,L^{-1}$. When determining analytes at very low concentrations so that the signal is right at the bottom of the calibration curve, it is important to ensure that the way that the calibration graph is plotted is appropriate. Calibration graphs can be plotted that have been forced through the origin, forced through the signal of the blank, not be forced through any point so that a line of best fit that takes into account all of the calibration points, *etc.* Careful consideration should be given when choosing which of the calibration options to use. Although negative concentrations can clearly not exist, it may be sufficient to simply state that those analytes are less than the limit of detection. Alternatively, if a customer requires positive numbers, then one of the alternative options for plotting the calibration may be used. For analytes that are hard to contaminate with, the blank value may be very close to zero anyway, in which case forcing through the blank or through the origin will not offer very different data. However, where analytes are easy to contaminate with, forcing through the origin would lead to an unrealistically high concentration to be provided. If the counts per second are lower than the zero standard, then either the zero standard has been contaminated (in which case another one can be prepared and the calibration adjusted accordingly) or there is some sort of physical effect that is suppressing the analyte signal in the samples compared with that in the standards. This can sometimes be corrected for by using an appropriate internal standard. Matrix matching standards and samples as far as possible also help minimise this problem.

The performance of an ICP–MS instrument should be checked before embarking on an analysis. It is routine for a performance report to be produced on a daily basis prior to use. As with ICP–OES, this involves the introduction of a specific standard available commercially or an identical one produced in-house at a hugely decreased price. In common with the performance reports obtained for ICP–OES, there is a measure of sensitivity where several elements covering the mass range will have to provide a minimum count rate. Similarly, there is also a measure of the precision. However, there is also a large number of other factors that the performance report checks. This includes checking that the mass spectrometer is registering a signal at the correct masses for certain analytes. This involves the determination of analytes such as Co, In and U, *i.e.* analytes towards the lower mass end, in the middle and towards the higher mass end of the spectrum. As well as ensuring that a signal is obtained for each of these analytes, it will also measure the resolution of the peaks for these signals to ensure that they are within set boundaries. For instance, it will look to ensure that each of the peaks is between 0.65 and 0.85 mass units wide. An under-resolved series of peaks may start to merge together, potentially leading to inter-element interferences. If a series of peaks are over-resolved, then the peaks obtained are too narrow and appear almost needle-like. This means that the mass spectrometer has a smaller target area to land on to register a signal. Therefore, tiny errors in where the mass spectrometer measures can lead to large uncertainty values and poor precision since it may land on top of the peak or it may measure half way up the peak. This is demonstrated in Figure 6.2. If the performance report indicates that sensitivity, precision or resolution are outside of acceptable parameters, then the analyst may tune the ion lenses to obtain a greater signal and adjust the resolution of the spectrometer. The latter may also affect the precision.

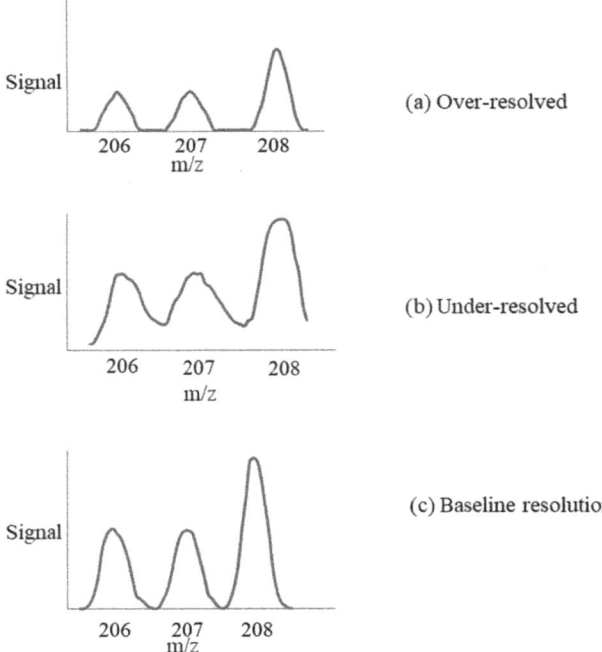

Figure 6.2 Over-resolved (a), under-resolved (b) and baseline-resolved (c) peaks for Pb.

A measure of doubly charged interferences is also made. The analyte detected is usually Ba. Again, a recommended maximum level of Ba^{2+} will be stated and the result clearly indicated. The extent of double ion formation can, in part, be controlled by adjusting the plasma operating conditions, *e.g.* the power and nebuliser gas flow rate.

Also included in the performance report is a measure of the count rate at a mass where there should be no signal, *e.g.* m/z 4.5. This will give an indication of the electronic noise associated with the instrument. This should typically be less than 0.5 counts per second. Often, a second mass towards the top of the mass range is also monitored. Since sensitivity increases with mass, the upper limit on this mass is more likely to be 2 or 3 counts per second. If the count rates are more than the maximum specified, then this could indicate something is wrong with the detector and that it may need replacing. Under such circumstances, it may be worth consulting a senior analyst/lab manager or contacting the manufacturer for advice.

Another important part of the performance report is to measure the extent of polyatomic interferences. This is often depicted as the $^{140}Ce^{16}O^{+}/^{140}Ce^{+}$ ratio which is often used as a proxy for all of the other polyatomic interferences. The amount of gas added to the collision cell can be adjusted and the ratio decreased. The ion lens settings may require adjusting for optimal signal. This can all be undertaken manually or using the "autotune" functions of the instrument. Often, a compromise between sensitivity and interference removal is required. Polyatomic interferences and how to overcome them will be discussed later in this chapter.

In common with the ICP–OES instruments, if a particular factor fails to achieve success during the performance report, the analyst must make a decision whether to continue regardless or to try and solve the problem. If a particular analyte fails a sensitivity check, *e.g.* only 78 000 counts per second are achieved rather than the 80 000 required, then this is likely to cause much less of a problem than if only 200 counts per second were achieved for all the analytes measured. Such a low signal would indicate a major flaw in the sample introduction system, the wrong solution is being nebulised, there is a blockage in the nebuliser, torch or cones, or wildly inappropriate lens settings are being used. An analyst should check the easy things first, *e.g.* sample introduction system and that the correct solution is being aspirated before fiddling with the ion lens system settings. A very minor drop in sensitivity for one analyte, especially when all other parameters meet specification, will lead to a calibration curve of marginally decreased sensitivity, but other than that, is unlikely to cause many problems. Unless the standard operating procedure or laboratory rules specifically state that all parameters must pass the performance report before an analysis can be undertaken, many analysts would continue without too many qualms. More consideration would be given to the level of polyatomic interferences. This could have a significant effect on the accuracy of data achieved, and so if this fails in the performance report, adjustment to the collision cell gas flow rate and ion lens settings may be required.

One aspect not often checked during a performance report is the position of the torch with respect to the sample cone orifice. If the cones have been removed to be cleaned or replaced, then there is a chance that they may not be placed back in exactly the same position on the interface. Even a slight deviation, *e.g.* by <0.5 mm can make a big difference to the sensitivity obtained. It is therefore necessary to optimise the *X*, *Y* and *Z* position of the torch with respect to the sample cone orifice to ensure maximum sensitivity.

One of the main problems with ICP–MS analyses is, as discussed previously, that of polyatomic interferences. In modern instrumentation, these can largely be overcome by using a suitable gas in the collision cell. A large number of studies have reported the use of numerous collision cell gases, because some gases work better for some analytes than others. If a suite of analytes is to be determined, then most laboratories use a gas that is a compromise, *i.e.* one that works reasonably well for most analytes, but may not be perfect for any. A good recent overview of interference types (including polyatomic), how to overcome them and the mechanistic aspects of how they are formed and removed was provided by Balaram.[9] A much older, but still very useful discussion of collision and reaction cells was presented by Koppenaal *et al.*[10] The paper by Koppenaal *et al.* discussed the early developments, design and applications of the cells. A third paper of relevance to this section was presented by Yamada.[11] This paper discussed the mechanism of kinetic energy discrimination (KED) and the mechanisms of the events occurring in both collision and reaction cells. Between them, these papers and the references therein will provide the reader with a good understanding of how collision and reaction cells work.

It should be noted that the parameters tested and the boundaries recommended for optimal performance vary significantly between a performance report for normal operation and another for operation using kinetic energy discrimination, *i.e.* the use of a cell. If an analyst attempts to pass a normal performance report when operating using a collision cell, it is very unlikely that many of the different parameters will reach the desired values. As noted earlier, the use of a collision gas adversely affects the sensitivity – especially at the lower end of the mass range. Therefore, sensitivity is likely to fail, possibly along with precision and resolution measurements. **Care should therefore be taken to ensure that the correct performance report template is used for different applications**.

Other problems may exist that a performance report does not check. Included in this number is ensuring that the detector voltage is on the "plateau region" of the response curve. As the detector ages, a higher voltage is required to obtain a useful signal. Therefore, the voltage required for a new detector to give a signal will be insufficient for one that is a year old. Instead, a signal of close to zero will be obtained. The analyst should also ensure that too high a voltage is not employed because that would age the detector prematurely decreasing its useful life. Although the results of the analysis may be unaffected if too high a voltage is employed, it should be noted that detectors are not cheap and that having to order one regularly is likely to lead to awkward questions from a laboratory manager. When adjusting the detector voltage, a standard should be introduced and the signal monitored. This will be observed to increase as the voltage increases and then reach a plateau where it does not increase further with increasing voltage. The voltage should then be reduced back to just above the point where the plateau in the signal begins.

As discussed in a previous section of this chapter, the linear range obtainable using an ICP–MS instrument can extend to over nine orders of magnitude. If fitted only with an electron multiplier detector, this would have to be operated in two different modes: analog and pulse count, where pulse count can be used up to approximately 3×10^6 cps and analog for up to 10^9 cps. The instrument is programmed to switch between the modes automatically. However, the two modes need to be cross-calibrated to ensure continuity of linearity; otherwise, there may be a "dog-leg" bend at the switch over point. If such a bend occurs, inaccuracy will occur. This cross-calibration should be undertaken periodically using a specific standard that is provided by the instrument manufacturers. It is also likely to be on the list of things to do during a preventative maintenance service by the manufacturers. If the laboratory undertakes the cross-calibration themselves, then a standard operating procedure will probably be provided. If not, then a junior analyst should seek advice or training from one that is more experienced.

In common with ICP–OES, if the torch is removed for cleaning, then it should be dried thoroughly before being re-installed. If any residual moisture is left in the torch, then when an attempt is made to re-light the plasma, there is a very good chance that a plasma will not form. Instead, an out-of-control discharge that cracks and pops will occur and, if unlucky, there is a chance that the torch will melt. This is obviously bad news because of the inconvenience and cost of re-placing the torch. However, a molten torch can also leave a white deposit of silica over the sampler cone and within the interface region which must also be cleaned before trying again. The torch may be dried with a tissue, but this is likely to leave areas that cannot be dried because they are inaccessible. A small squirt of acetone through each of the three tubes comprising the torch dries it more efficiently. If the gas flow through the torch time is extended without attempting to light the plasma, then this ensures that all traces of the acetone are removed.

Some nebuliser/spray chamber assemblies require a very precise coupling. For instance, if the nebuliser is placed into the spray chamber up to the gas inlet, this may lead to a significant decrease in sensitivity compared with when it is withdrawn just 2–3 mm further out. Manufacturers often provide a convenient spacer that can be placed on the nebuliser enabling it to achieve optimal coupling position. It is therefore not just the gas flow that needs optimising when tuning the instrument.

There are some other potential problems/errors that can occur under certain circumstances. An example is when organic solvents are introduced. These have numerous problems associated with them including increased vapour pressure, often lower boiling points and lower surface tension; all of which will contribute to an increased sample transport efficiency to the plasma and hence increased plasma loading. They also have a larger dissociation energy and hence will drain power out of the plasma, leaving less power to atomize and ionize the analytes. They also decompose in the plasma producing large amounts of soot which, in turn, starts to block interface cones and to coat ion lens systems. The soot will cause serious sensitivity drift until blockage is complete at which point no signal will be observed. A way of overcoming this is to introduce oxygen to the plasma and most manufacturers provide a gas inlet for this purpose. However, there is a fine line between adding sufficient oxygen and adding too much. An excess will have a seriously deleterious effect on the sampler cone, with the orifice becoming enlarged; possibly to the extent where a sufficiently good vacuum in the interface region cannot be obtained. A new cone would then be required. This process can be rapid if far too much oxygen is added. The use of a platinum-tipped sampler cone as opposed to a standard nickel one can diminish the adverse effects of oxygen addition significantly since it is much more robust. However, the downside is cost, with a platinum cone being far more expensive than a standard one. An overview on the introduction of organic solvents to ICP–MS instruments was published in 2015 by Leclercq *et al.*[12] This review highlighted many of the applications that had been undertaken as well as the different instrumentation used for the introduction. These can be assorted desolvating types of spray chambers (including those that operate at reduced temperature), very low flow nebulisers, direct injection nebulisers, desolvation membranes and other, lesser used, methods of sample introduction, *e.g.* electrothermal vaporization. A useful table that gives the chemical compatibility of components to assorted solvents was also given. This point should not be overlooked since standard peristaltic pump tubing is not resistant to many organic solvents and will disintegrate quite rapidly. This runs the potential risks of delayed analysis, lost sample and damage to the peristaltic pump. Whenever organic solvents are used, it is therefore advisable to use specialist organic solvent-resistant pump tubing.

A further problem with the introduction of organic solvents is the instability of the plasma that they cause. Very high reflected powers and possible plasma extinction occur when some solvents are introduced. Careful optimisation of the operating conditions is therefore required, with many workers opting to use slightly higher forward powers than normal (*i.e.* in excess of 1400 W). The paper by Leclercq *et al.* discussed the general trends with regard to optimising the assorted gas flow rates that form the plasma, viewing heights (sampling depths for ICP–MS instruments), *etc.* Also highlighted was the requirement for matrix matching and the applications for isotope dilution. The information given in this paper, and the 305 references therein, should provide a very helpful guide to those new to the area. Although most laboratories do not require the introduction of organic solvents, those working in the petrochemical industry might. Similarly, some speciation analyses may also rely on a chromatographic system that has an organic solvent as part of the eluent.

Many instrument manufacturers provide information on how to introduce organic solvents to a plasma while causing minimal disruption. An example is this is provided by Agilent.[13]

Other sample types can also be problematic. For example, samples that have very high concentrations of sulfates, *e.g.* digests of plastics that are often dissolved using a sulfuric/nitric acid mixture, can be equally troublesome. The nickel cones are very susceptible to damage caused by sulfates (and to a lesser extent, phosphates). Therefore, if a series of sulfuric acid digests is to be analysed, the use of a platinum-based sampler cone is recommended.

As already discussed, a problem can be experienced when the orifice of the sampler cone becomes enlarged to the extent that the interface region cannot reach the desired vacuum. This occurs naturally with extended use of cones and need not require one single event such as those discussed above. When enlargement of the orifice does occur, the plasma will probably light, but when the vacuum pump to the interface turns on, it will not be able to reach the interlock value enabling the slide valve to open. The plasma will either then turn off, or the instrument should show a warning message. The only useful way to overcome this is to install a new cone.

If an isotopic analysis is to be undertaken, *e.g.* for an isotope dilution analysis or if isotopic ratios are to be determined then several potential errors may occur. Isotope dilution analysis is generally regarded as being an extremely accurate method of analysis. This is because it can correct for analyte loss and instrumental drift problems as well as some other problems. For isotope dilution experiments to occur, then the analyte must have more than one stable isotope. A first problem often encountered is locating isotopically altered elements that are well-characterised and are of an affordable price. The exact composition of the spike material must be known for all further calculations. If it is unknown or inadvertently contaminated, severely inaccurate data will be produced. A tutorial to isotopic analysis was presented by Alvarez-Penanes *et al.*[14] Prior to any instrumentation being used, there are several requirements necessary to obtain accurate data. The spike must be allowed to equilibrate with the sample for a sensible length of time. The length of time required depends on numerous factors, *e.g.* the sample type, the analyte, whether "total" analyte is required or whether it is a speciation analysis being undertaken, *etc.* Liquid samples require no real equilibration time because the spike may simply be stirred into the sample to ensure homogeneity. For the determination of "total" analyte concentrations of an organic material, *e.g.* a tissue sample, vegetation, blood, *etc.* a long equilibration time may not be necessary because the sample can be acid digested completely. However, if a speciation analysis is to be performed, then a much milder extraction protocol is likely to be required. This means that isotopically enriched species rather than ions need to be spiked into the sample. This can be problematic in that relatively few isotopically labelled species exist and hence many have to be prepared in-house. The cost of this is prohibitively expensive for many laboratories because on top of the cost of the isotopically enriched analyte, if different species of an analyte need to be prepared, a synthetic chemist may have to be employed.

When undertaking an isotopically enriched speciation analysis, there is clearly no point in spiking a sample and then immediately extracting it because that would not enable it to become realistically embedded in the material. Instead, it would simply be washed off. This would potentially lead to significant error. "Soaking in" or equilibration times of several hours or even days may be required before the extraction procedure should start.

Assuming the sample has been prepared and extracted in a suitable way, then the problems associated with the instrumental analysis can be addressed. For the most accurate and precise isotopic data to be obtained, a multicollector instrument would be used so that all of the isotopes may be recorded simultaneously. Quadrupole-based instruments, although very rapid, are still sequential in nature. This inherently leads to poorer precision and potential error, although many laboratories manage to obtain adequate data using them. Spectroscopic interferences should be minimised by ensuring that all isotopes are measured free from isobaric and polyatomic interferences. This may mean using collision/reaction cells or using high resolution ICP–MS.

Once an analysis is underway, the analyst should be wary of the mass bias effect. This is dependent on the sample matrix but may also vary with time and even fluctuate randomly. Although a detailed description is not suitable for this script, it is necessary to know that the mass bias effect is where the lighter ions are discriminated against because of the space charge effect. These therefore have a slightly lower sensitivity than higher mass isotopes – even for the same analyte. It is therefore advisable to run samples and isotopic standards and another element that is close in mass to the isotopes of interest at regular intervals to monitor and correct for the effects. Some laboratories go for a "bracketing" approach, where the order of analysis will be sample, standard, sample, standard, *etc.* The sample bracketing approach to the analysis inevitably leads to a significant lengthening of the time required for the analysis.

6.4 Hints and Tips

As for all analytical techniques, it is good practise to use certified reference materials (or in-house reference materials if CRMs are unavailable) to ensure the accuracy of the analysis. Periodic check standards to monitor signal drift are also recommended. As with all analytical techniques, if a standard or reference method is available that discusses the analysis of a particular sample type, it is easier to follow that rather than try and develop a method from scratch. Similarly, the instrument manufacturers have experienced application scientists who may be able to advise. Sometimes manufacturers produce technical notes that may also be of interest or assistance.

One of the main problems associated with ICP–MS analysis is that of contamination. It is therefore recommended that if undertaking ultra-trace analysis, the purest of chemicals be used (or procedures such as sub-boiling distillation be used for purification). In addition, as much sample manipulation as possible should be performed in clean rooms and/or laminar flow hoods. It is important to note that acid digestions should not be performed in laminar flow hoods as the fumes would be blown into the laboratory.

As with ICP–OES, the use of an internal standard (or standards) is also recommended. The internal standard should not be in the sample at a significant concentration naturally. It should be as close to the analyte in mass and ionization potential as possible. Therefore, if a range of analytes covering the entire mass range is to be determined, then several internal standards may be required. Examples could include Co, In, Ir and possibly Bi. The Co could be used for analytes at the lower mass, In for those in the middle mass range and either Ir or Bi at the top. It is possible to use a mixture of internal standards for an analyte, *e.g.* both In and Ir could be used for analytes whose mass is between them, *e.g.* the rare earth elements. The instrument's software should allow the analyst to pick which internal standard(s) to use for which analytes. It should also be possible to change them before, during or after the analysis to obtain the best results (assuming all of the internal standards have been measured). It should be noted that the list of internal standards given above is not exhaustive and the examples given are useful for an acid medium. If enzyme extracts are to be analysed, then these internal standards may not be stable at a neutral pH and could therefore yield very dubious data. It may be necessary to use an element such as Cs that is stable at most pH values. The choice of internal standard is also dependent on the sample matrix. If working in a semiconductor factory, it may be unwise to use indium which could be a major matrix component. As specified previously, the internal standard should not be present at an appreciable concentration in the sample naturally. Therefore if, in a soil digest, the In concentration naturally is $10 \, \mu g \, L^{-1}$, then a dilution by a factor of at least 100 is required before adding a spike of In internal standard of $10 \, \mu g \, L^{-1}$. This would lead to an error of only 1% rather than the original error of 100%. Had no dilution been made, then the analyte concentrations would have been underestimated by a factor of two. The alternative would be to use a different internal standard that is not present naturally or, if it is, it is present at an insignificant concentration, *e.g.* $<0.1 \, \mu g \, L^{-1}$.

Matrix-matching standards to samples is usually a good idea. Some analytes require the presence of chloride to be solubilised from the sample and to stay stable in solution. The platinum group elements belong to this group. However, the presence of chloride can be problematic for elements such as Ag and Pb, which can precipitate. This is not normally a problem at the concentrations analysed using ICP–MS, but the reader should be aware that intermediate standards, *e.g.* $10 \, mg \, L^{-1}$ that are used to prepare working ICP–MS standards could precipitate.

The internal standard can be administered in more than one way. It is possible to add it to every blank, sample and standard using a pipette. Assuming it is mixed to ensure homogeneity before the analysis, this can be very accurate. However, it does lead to the possibility of samples being accidentally left out or double-spiked with internal standards if the analyst loses concentration during the spiking process, *e.g.* if somebody comes to speak to them. Errors may also be made if some volumetric flasks are of different sizes, *e.g.* a 50 mL one is used instead of 25 mL. In this instance, if the same volume of internal standard is added to the 50 mL flask as the 25 mL ones, the concentration of the internal standard would be half of what it should be. Therefore, the instrument software will artificially double the concentrations for those analytes that use that internal standard. The alternative method of administering the internal standard is to merge a solution of it with the sample *via* a T- or Y-piece prior to the nebuliser. This method assumes adequate mixing occurs and that the peristaltic pump tubing of both streams does not begin to wear.

Different geometry or bore of injectors for samples high in dissolved or suspended solids or for organic solvents are available.

If there are significant concentrations of suspended and/or dissolved solids present in a sample, *e.g.* blood, then torch injectors that have constrictions are likely to block. Instead, an injector with a gradually tapering bore would be better. Similarly, the bore of injectors used when organic solvents are to be introduced tends to be wider than those for aqueous-based samples. Most manufacturers will provide a torch assembly to be used for aqueous-based samples and another for organic-based ones. Similarly, if samples have been dissolved using hydrofluoric acid, the standard quartz torches (plus the glass nebulisers and spray chambers) would be severely damaged if all of the acid had not been removed. As little as 1% remaining can cause damage. A ceramic-based torch and/or a perfluoroalkoxy alkane-based nebuliser/spray chamber assembly is often provided to prevent this damage.

The nebuliser and spray chamber types are also important for some applications. In general, ICP–MS is the analytical technique of choice as a detector for chromatographic separations during speciation analyses. Since liquid chromatography peaks can be fairly broad (depending on the column dimensions, *etc.*), then they should not be broadened further because there is a risk that they would start to coalesce. It is therefore important to have a nebuliser/spray chamber assembly that is rapid clearing to prevent this. A pneumatic nebuliser and cyclone spray chamber would be better choices than a high solids nebuliser and a Scott double pass spray chamber. The Scott double pass spray chamber is excellent at dampening noise arising from peristaltic pumps because it has a large volume and surface area. However, these attributes lead to severe broadening of chromatographic peaks because of the long washout period required.

Another aspect to consider is the mobile phase of liquid chromatography. It is possible that this may cause interferences. For instance, if a gradient or step gradient of potassium sulfate is used as a mobile phase during arsenic speciation, there is a good chance that the polyatomic ion $^{39}K^{36}Ar^+$ would be formed which could interfere during the determination of $^{75}As^+$. Although the collision cell could overcome this quite readily, it is likely to lead to an unwanted decrease in sensitivity. It is therefore better to change the mobile phase slightly to something like sodium sulfate so that the interference is avoided completely.

Speciation analyses using chromatographic separations require the use of the time resolved software that is in-built to the instruments' software. Here, the analyte(s) of interest can be separated and quantified. A chromatogram of standards of the different species can be obtained and then the program is instructed to integrate over specific time periods correlating to those times where the different species elute. The computer can then use the area under the standard peaks and under the sample peaks to calculate concentrations. Since some chromatographic separations can take several minutes to be completed, the process is not fast. However, by programming the instrument to automatically calculate the concentrations, some time is saved when compared with exporting the data into a software package such as Excel and then integrating manually. When setting up the software to measure specified time regions of the chromatograms, it is obviously necessary to ensure that the sample is injected and the analysis started simultaneously. If not, the regions integrated may not coincide with the analyte peaks.

Some speciation analyses use gas chromatography to separate the analytes. This is a less straightforward coupling than with liquid chromatography which often uses a simple connection from the outlet of the chromatographic column to the nebuliser (although the length of this should be minimised to prevent peak broadening). For gas chromatography, the analytes elute from the hot column and must be kept in a gaseous state, *i.e.* kept above their boiling points, right into the torch. If there are any cool spots, then the analytes will condense and be lost analytically. The heated transfer line is placed up to the tip of the injector of the torch to avoid this. Although the preparation of heated transfer lines in-house is possible, it can be a lengthy process with many designs failing. It is therefore wise to use the specialist interfaces supplied by the instrument manufacturers. In comparison, these are relatively efficient and straightforward to install. An annual review of speciation methods utilising atomic spectrometry (not just ICP–MS) is published in the *Journal of Analytical Atomic Spectrometry*.[15]

Another application area that is increasing in popularity is the analysis of nanoparticles. Nanoparticles can be industrially machined, *e.g.* for catalysts, may be the result of decomposition, *e.g.* the formation of nanoplastic particles from larger materials, food additives (E551), or can even be single cells from biological materials. The experiments may be a simple characterisation, *i.e.* what the particles are made of, a more complex toxicology study or a characterisation in terms of both composition and particle size distribution. A recent tutorial review discussing the analysis of nanoparticles was presented by Meermann and Nischwitz.[16] Some engineered nanoparticles have a tendency to agglomerate and so a dispersant is often employed along with either ultrasonication and/or vortex mixing to try to disaggregate any clusters. For particle size characterisation, the samples may be separated using field flow fractionation and then detected using ICP–MS. The paper by Meermann and Nischwitz also discussed the different modes of field flow fractionation and hydrodynamic chromatography coupled with ICP–MS. Alternatively, the increasingly popular technique of single particle analysis may be employed. A good recent overview of single particle analysis was presented by Resano *et al.*[17] The overview contained 216 references and covered topics such as the theory, appropriate certified reference materials, methods of improving transport efficiency, calculation speed, minimising matrix effects and spectral overlap as well as the analysis of single cells. In short, for optimal performance, a low concentration of particles and a rapid acquisition speed are required to prevent more than one particle being detected simultaneously. A high efficiency sample introduction system is also required for a greater signal. For a conventional analysis, a dwell time of approximately 10 ms is often used. However, for single particle analysis, it is normal to use much shorter times, *e.g.* 1 ms or even less. Again, the time resolved software is required for the analysis. This time possibly thousands of spikes are obtained with each spike corresponding to the detection of a single particle and the height of the spike corresponding to the particle size. In the early stages of single particle analysis, the analyst would have to undertake laborious and challenging calculations to try and determine the particle size distribution and number of particles present. Nowadays, manufacturers have provided software packages to enable this to be completed more readily. Alternatively, software for interpreting single particle analysis of nanoparticles that is independent from manufacturers has also been developed.[18] For any analyst entering this topic area, it is worth noting that two limits of detection are possible: one for the lowest concentration and the other for the smallest particle size determinable. Most software packages assume that the particles are spherical. In practise, this is obviously not necessarily the case.

6.5 Triple Quadrupole/Reaction Cell Instruments

In general, this class of instrument shares many of the same advantages and disadvantages as single quadrupole ones. However, they are approximately 30–50% more expensive. They do, however, have the large advantage of offering a further mode of analysis. As well as the normal and kinetic energy discrimination modes offered by single quadrupole instruments, they also offer the possibility of chemical reactions. This is where a reaction gas is introduced to the cell where it reacts with the analyte effectively moving it away from its own mass. For example, during the determination of P, oxygen is often added so instead of measuring the P at m/z 31, it is measured as $^{31}P^{16}O^+$ at m/z 47. This has the advantage of moving it away from the polyatomic interferences exerted at m/z 31 (*e.g.* $^{17}O^{14}N^+$, $^{16}O^{15}N^+$, $^{30}Si^1H^+$, *etc.*) whilst simultaneously increasing sensitivity because of the higher mass being monitored and improving limits of detection. Similar reactions may be used to determine other analytes, *e.g.* S. Other reaction gases may also be used. One of the other favoured gases is ammonia, which has been used to react with Ti. An example of how the reaction cell works is shown in Figure 6.3. All analyte ions and surviving molecules enter the first quadrupole. This then isolates the species that exist at the relevant mass. In the case above, this is $^{87}Sr^+$ as well as $^{87}Rb^+$. All other masses are lost from the system. The ions at m/z 87 then enter the reaction cell (Q2), where they are reacted with a gas. When oxygen is used, $^{87}Sr^+$ readily forms the oxide at m/z 103, whereas $^{87}Rb^+$ does not. The third quadrupole then passes $^{87}Sr^{16}O^+$, now interference free, to the detector.

Many instrument manufacturers produce a reaction cookbook, *i.e.* a handbook that recommends which gas to use for which analyte. Alternatively, these recommendations may also be part of the instrumental software. Good and relatively recent tutorial reviews of triple quadrupole (tandem) ICP–MS instruments have been provided by Bolea-Fernandez *et al.*[19] and Balcaen *et al.*[20] Both give a good overview of the different modes of analysis and tabulate some of the applications that had been undertaken. The paper by McCurdy and Yamanaka, discussed previously, is also a good read for an analyst presented with a new instrument.[3]

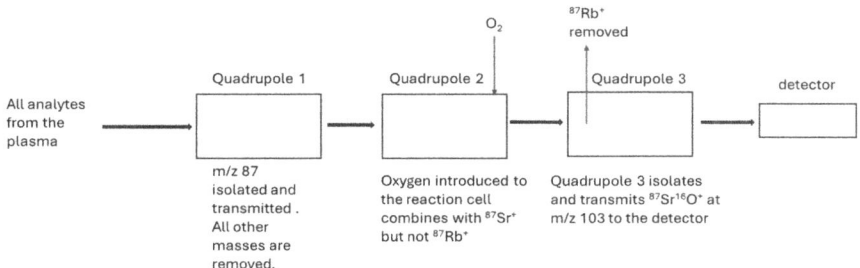

Figure 6.3 An example of how a reaction cell works.

As discussed, this type of instrument has the ability to determine analytes in the normal mode, kinetic energy discrimination mode (collision cell) and the reaction cell (also known as off mass analysis). All three modes of analysis may be used during the same analytical experiment, albeit at the cost of increased time. This is because the collision/reaction cell needs to be evacuated and/or the gas changed between analytes. For the very light elements, e.g. ^{7}Li, ^{9}Be and ^{10}B, normal mode single quadrupole analysis is usually recommended. These analytes have relatively poor sensitivity because of their low mass and the use of kinetic energy discrimination would likely lead to a signal loss of >90%.

The manufacturers' recommendations for the reaction gas should normally be applied to the analysis. If a laboratory does attempt to use a different gas, the analyst should first ensure that it does not react with or otherwise degrade the components of their instrument. Manufacturers usually provide two or possibly three gas inlets so that alternatives may be used. They would also be able to advise whether or not the instrument they provided will be damaged by another gas.

6.6 Sector/Multicollector Instruments

Numerous versions of these instruments exist including magnetic sector instruments, electrostatic sector instruments, single collector instruments and multicollector instruments. A schematic diagram of a sector field instrument is depicted in Figure 6.4. The sample introduction system has not been included but is often similar to that for any other ICP instrument. These instruments are significantly more expensive than the other types, with a single collector sector instrument being 2–3 times more expensive than a triple quadrupole instrument and a multicollector being 3–4 times. This is obviously prohibitively expensive for many laboratories. They do have several advantages including being extremely sensitive (even more so than most quadrupole-based instruments) and having a far higher resolving power in that they can resolve down to well below 0.01 mass units compared with unit mass for quadrupole-based instruments. Hence, they suffer fewer polyatomic and isobaric interferences since the signals from the analytes are more easily separated from those of the concomitant ions. Detection limits are quoted to be at the fg L^{-1} level, although pg L^{-1} would be more normal. The linear range can also span nine orders of magnitude (from one count per second to 6 giga counts per second). Another advantage is that they do not require

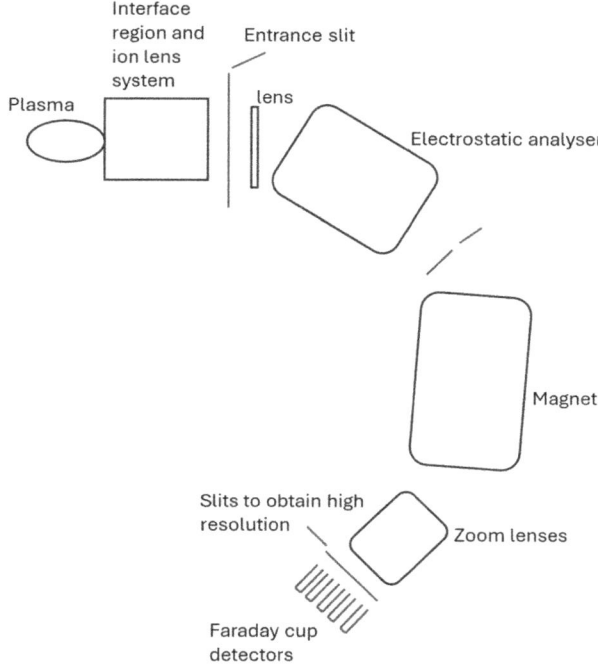

Figure 6.4 A schematic diagram of a multi-collector ICP–MS instrument.

any reaction gases to remove interferences. The downside is that the instruments often need to be installed in special clean room facilities (hence increasing the cost further). If high resolution is required to separate an analyte from an interference that is very close in mass, a decrease in sensitivity results. Although the decrease in sensitivity is unfortunate, it is a worthwhile sacrifice to obtain an interference-free analysis. Fortunately, many of the interferences that may be overcome require only medium resolution leading to only a marginal decrease in sensitivity. For those analytes requiring high resolution, *e.g.* when $^{75}As^+$ (atomic mass 74.9216) is to be resolved from $^{40}Ar^{35}Cl^+$ (molecular mass 74.9169) a significant decrease in sensitivity would be observed. The resolution should be such that a flat-topped peak is obtained. The flat-topped peak enables the best accuracy and precision to be obtained compared with the rounded peaks obtained in quadrupole-based instruments. This is because any lateral movement in the mass that is measured will have an inconsequential effect on the signal. A good tutorial overview of sector field measurements of isotopic ratios was presented by Baxter *et al.*[21] The review discussed noise sources that affect the precision, problems that affect accuracy (mass discrimination), methods to overcome the problems (normalisation, sample standard bracketing) and the effects of detector dead time.

Multicollector instruments have numerous detectors and often different types of detectors in the same instrument. One commercial instrument has nine Faraday cup detectors, two compact Discrete Dynode Multipliers for analytes at higher concentrations and an axial discrete dynode secondary electron multiplier. Other instruments have 16 or more detectors. In general, the secondary electron multipliers are used for the analytes yielding low signal, *i.e.* are present at low concentration. The mechanisms by which all of these detectors work are well out of the scope of this text, but may be found in other publications.[1] Many of the detectors are usually moveable to ensure that different ion masses may be captured. It should be noted that the number of analyte ions that can be detected simultaneously will depend on the number of detectors present. An analyst hoping to detect 60 different m/z simultaneously is therefore likely to be disappointed. Another limiting factor of this type of instrument is the maximum dispersion available. This is the mass range detectable as a percentage of the middle mass being determined. If the maximum dispersion is 20%, then this means that there will be 10% on either side of the middle mass. The range is obviously greater at higher mass, *e.g.* if the middle mass is m/z 210, then the range will be 210 ± 21, *i.e.* 189–231 compared with when the middle mass is 30 (the range will be only 27–33). Depending on the application, it may be necessary to run the samples more than once so that all isotopes can be determined. The analyst should therefore be aware that being billed as a simultaneous instrument does not necessarily mean that it can detect all analytes simultaneously.

In practise, many applications that utilise such an instrument do require a very limited number of isotopes to be determined, *e.g.* Pb isotope pairs (plus a mass bias correction isotope), or Sr isotopes, *etc.* To obtain optimal sensitivity and interference removal, many laboratories undertake extensive sample clean up procedures so that only the isotopes of interest are left, *i.e.* isobaric interferences are removed through the judicious use of resins. Similarly, digestion acids or other materials present as concomitants that are likely to lead to polyatomic interferences are also removed. In this way, higher resolution is not required leading to increased sensitivity.

As discussed previously, different instruments have different applications. If a straightforward reduction of interferences is required on one or several analytes, then a single detector instrument would suffice. If isotope ratio measurements are to be performed then a multicollector instrument will provide the best accuracy and precision.

6.7 Data Analysis

The data analysis requirements for ICP–MS are very similar to those of ICP–OES. All replicate measurements of a sample should be inspected to ensure that they are all similar, *i.e.* that the sample had arrived to the plasma in time to be measured (if not, the first replicate at least may be lower than expected) or had not run out before measurement was complete (the last replicate would be lower). Routine check standards every 10 samples would help identify any drift. This is, however, usually undertaken and corrected for during the analysis. If a particular internal standard is noticed to be drifting badly but the other ones are remaining stable, it could be that it is not stable at the pH at which the samples are present. If this is the case, then it may well not be correcting for any drift the analytes that use it as an internal standard are experiencing. If this is the case, then the analyte concentrations should be re-calculated using a different internal standard. Comparison of experimental data with certified or expected values for reference material as well as monitoring the check standards will identify whether or not this has rectified the problem. Similarly, if a sample has double the concentration of an internal standard, it may have been double spiked. Again, removal of the internal standard or choosing a different one may rectify the problem. If the internal standards remain fairly stable throughout an analysis with the exception of just one or two samples, this is usually a sign that an error was made when spiking.

As always, if the experimental data for a certified reference material do not match the certified values but there appear to be no problems during the analysis, it could be that an error was made during the calculation and preparation of the standard concentrations. These should be checked. If an error has occurred, it is usually possible to re-program the ICP–MS instrumental software with the correct concentrations or to re-prepare the standards and re-run them on the instrument to re-calibrate it. If the problem is noticed early in the analysis, *i.e.* if a certified material sample is run shortly after the calibration is completed, then it is easier to correct for this type of error. Other than not noticing a problem and reporting wholly incorrect data to a client, the worst case scenario would be to notice that something had gone wrong after the analysis is complete. Correcting the problem a day or two after the experiment is complete may well involve re-running the entire experiment again. If another analyst has used the instrument in the interim, they may have changed the operating conditions and ion lens settings or they may have used samples with high solids and have had to remove and clean the cones. All of these mean that the sensitivity cannot be relied on to be the same as when the samples were analysed initially necessitating a full re-analysis. If sample volume had been limited to start with, then this could be seriously problematic. It is always wise for an analyst to watch the calibration as it is formed and then inspect the results for one or more certified materials to ensure that they make sense before continuing with the experiment. This is true especially if an auto-sampler is used because it is tempting to just hit the "go" button and rush off to do something else.

References

1. R. Thomas, *Practical Guide to ICP–MS and Other Atomic Spectroscopy Techniques, A Tutorial for Beginners*, CRC Press, Taylor & Francis, 4th edn, 2023, ISBN 10:1032035021.
2. K. L. Linge and K. E. Jarvis, *Geostand. Geoanal. Res.*, 2009, **33**, 445–467.
3. E. McCurdy and M. Yamanaka, *Spectroscopy*, 2022, **37**, 42–45.
4. B. Woods and E. McCurdy, *Spectroscopy*, 2022, **37**, 54–56.
5. E. H. Evans and J. Giglio, *J. Anal. At. Spectrom.*, 1993, **8**, 1–18.
6. T. May and R. H. Wiedmeyer, "A Table of Polyatomic Interferences in ICP–MS" (perkinelmer.com).
7. https://www.agilent.com/cs/library/applications/application_double_ion_icp-ms_8900_5994-1155en_us_agilent.pdf.
8. C. Duyck, R. R. A. Peixoto, A. A. Rocha, H. G. D. Severino, P. V. Oliveira, R. Damasceno and R. Lorencatto, *J. Anal. At. Spectrom.*, 2022, **37**, 474–496.
9. V. Balaram, *Rapid Comms. Mass Spectrom.*, 2021, **35**, e9065.
10. D. W. Koppenaal, G. C. Eiden and C. J. Barinaga, *J. Anal. At. Spectrom.*, 2004, **19**, 561–570.
11. N. Yamada, *Spectrochim. Acta*, 2015, **110**, 31–44.
12. A. Leclercq, A. Nonell, J. L. Todoli Torro, C. Bresson, L. Vio, T. Vercouter and F. Chartier, *Anal. Chim. Acta*, 2015, **885**, 57–91.
13. Robust Plasma Conditions for the Analysis of Trace Metals in Organic Samples, https://www.agilent.com/cs/library/slidepresentation/Public/Organics%20Analysis.pdf.
14. P. Alvarez-Penanez, A. Reguera-Galan, G. Huelga-Suarez, A. Rodriguez-Castrillon, M. Moldovan and J. I. Garcia-Alonso, *J. Anal. At. Spectrom.*, 2022, **37**, 701–726.
15. R. Clough, C. F. Harrington, S. J. Hill, Y. Madrid and J. F. Tyson, *J. Anal. At. Spectrom.*, 2023, **38**, 1339–1371.
16. B. Meermann and V. Nischwitz, *J. Anal. At. Spectrom.*, 2018, **33**, 1432–1468.
17. M. Resano, M. Aramendia, E. Garcia-Ruiz, A. Bazo, E. Bolea-Fernandez and F. Vanhaecke, *Chem. Sci.*, 2022, **13**, 4436–4473.

18. T. E. Lockwood, R. G. de Vega and D. Clases, *J. Anal. At. Spectrom.*, 2021, **36**, 2536–2544.
19. E. Bolea-Fernandez, L. Balcaen, M. Resano and F. Vanhaecke, *J. Anal. At. Spectrom.*, 2017, **32**, 1660–1679.
20. L. Balcaen, E. Bolea-Fernandez, M. Resano and F. Vanhaecke, *Anal. Chim. Acta*, 2015, **894**, 7–19.
21. D. C. Baxter, I. Rodushkin and E. Engstrom, *J. Anal. At. Spectrom.*, 2012, **27**, 1355–1381.

7 Laser Induced Breakdown Spectrometry

7.1 Introduction

Laser induced breakdown spectrometry (LIBS) is an increasingly popular method. It has now existed for over 50 years, but its popularity has increased most rapidly over the last 20 years. It is a light emission-based technique and the physics behind the detection is the same as for ICP–OES. As is the case throughout this book, the detailed theory will not be given. However, a brief summary is required. Instead of using a flame or a high energy plasma to excite the analyte molecules/atoms/ions in a sample, a laser is used. At the focal point of the laser, the temperature may reach between 20 000 and 40 000 °C, depending on the power and wavelength of the laser used and on the physical properties of the sample. Some versions use a single laser, others use one laser but apply a double pulse; one to vaporize and excite the sample and the next to further excite the vapour cloud. Other versions use two different lasers to obtain a double pulse. During the excitation of the atoms/ions, electrons are promoted to higher energy levels before relaxing back to their normal state, releasing energy in the form of light that is specific to each individual analyte. The amount of light emitted is proportional to the concentration of that analyte. A good recent overview of the assorted techniques was presented by Wang et al.[1] As well as discussing the different incarnations of LIBS, it also describes the

Practical and Technical Guides for Laboratory-based Chemists No. 1
Atomic Spectrometric Methods of Analysis
By Andrew Fisher
© Andrew Fisher 2025
Published by the Royal Society of Chemistry, www.rsc.org

assorted approaches to calibration and data interpretation. This publication would be a good place to start for any newcomer to the technique. Another review, containing 210 references, that highlighted the developments of LIBS in recent years was presented by Guo *et al.*[2] Many papers published have described the use of instruments made in-house. However, nowadays there are commercial instruments including handheld/portable ones. These have been used in an assortment of industries including the analysis of alloys, recycling of polymers, electronic materials, fuels and environmental analysis. A good recent review of both portable and handheld LIBS instruments containing 121 references was presented by Senesi *et al.*[3] The applications and characteristics of the instruments were presented including an extremely helpful table listing the characteristics of currently available handheld LIBS spectrometers. Other sections of the review cover applications including those from archaeology, clinical, industrial and the nuclear fields. Overall, LIBS is generally regarded as having the potential to become one of the most common methods of analysis. A good tutorial for the non-expert is available at the website: https://www.nitonuk.co.uk/wp-content/uploads/eBook-LIBS-Technology-for-Non-Scientists.pdf.[4] Numerous other reviews have been published recently that focus on different application areas including environmental systems,[5] cultural heritage and archaeology,[6] agriculture and food,[7,8] explosive residues,[9,10] clinical samples,[11] industrial samples[12] and geotechnical engineering.[13] Other, more specialised recent reviews also exist including ones for the analysis of coal[14] and plastics.[15] A review of portable instrumentation (including but not confined to LIBS) was presented by Crocombe.[16] The Atomic Spectrometry Updates published in the *Journal of Analytical Atomic Spectrometry* would also be good sources of information. The reader may understand from the extremely high volume of recent review papers that LIBS analysis is becoming increasingly popular. Although it does not provide the low limits of detection to replace extremely sensitive techniques such as ICP–MS, it will certainly compete very strongly with many of the others, *e.g.* flame AAS, XRF and ICP–OES. Another very recent and welcome addition to the literature has been presented by Singh *et al.*[17] who have published a two-volume book with individual chapters being provided by experts in the field. The contents of the volumes included the fundamental theory, multiple-pulse LIBS, standoff developments, nanoparticle enhancement, underwater applications, calibration strategies, chemometric analysis of spectra and numerous applications.

7.2   Advantages and Disadvantages of LIBS

The advantages and disadvantages of LIBS are summarised in Table 7.1 but are discussed in much greater detail in the text.

7.2.1   Advantages

Simultaneous multi-element determination is possible using many LIBS systems. This is especially true when using a charge coupled device (CCD) or intensified CCD detector. Compared with many techniques, it is also relatively inexpensive. This is especially true for instruments made in-house, where individual components may often be found lying around the lab or in drawers. The hand-held instruments are also significantly cheaper than instruments for some of the other techniques.

Each analysis takes fractions of a second, although more than one analysis is usually undertaken per sample. For double pulse LIBS, the two laser shots may only be a few microseconds apart, but the light detection, computing and interpretation take longer. In addition, rasters across the surface of the sample can be undertaken to test sample homogeneity and measurements at different depths may be made by repeated laser irradiation of the same spot. A recent review of the literature regarding the use of LIBS for mapping was presented by Limbeck *et al.*[18] The mapping required for a complete analysis of

Table 7.1   Advantages and disadvantages of LIBS.

Advantages	Disadvantages
Simultaneous multi-element determination	Calibration is not easy
Little sample preparation required	Optimisation of operating parameters critical to performance
Rapid analysis	Not usually as sensitive as ICP–OES or XRF
Standoff analysis capability	Not good with rough or uneven surfaces
On-line analysis capability	Ionising radiation from lasers is present
Many instruments made in-house	Self-absorption problems can be severe
May be applied to solid, liquid and gaseous samples	Precision is often poorer than other techniques
Relatively cheap	
Enables spatial information to be obtained	
Relatively non-destructive	
Yields quantitative and qualitative data	

cultural heritage samples, bio-imaging and materials science were discussed.

Another major advantage of LIBS is that it needs negligible sample preparation. Since this is often the Achilles heel of any analysis in that it can lead to contamination and analyte loss, is often time-consuming and frequently requires unpleasant chemicals for sample dissolution, the fact that LIBS does not require it is a huge bonus. That having been said, extraneous materials, *e.g.* rust on a metallic artefact, soil or sediment on a heritage sample, *etc.* should be removed carefully prior to analysis to ensure that a surface representative of the sample is analysed.

The ability to undertake standoff analysis is another huge benefit. Some sample types are dangerous, *e.g.* explosives and components of a nuclear reactor. The ability to analyse these from a distance enables the analyst to remain safe. The way this is undertaken depends on the sample type. It is possible to fire a laser through a fiber optic so that the sample is vaporized and excited. The light may then be collected and transmitted *via* a second fiber optic back to a spectrometer and a detector. Such a system may be applied to the analysis of nuclear materials by passing the fiber optics through a window directly into the reactor. By passing the laser light through the optic and the transmitted light back through the window, the analyst is kept safe from the obviously dangerous materials under analysis. It has the added bonus of enabling the analysis to be undertaken without the necessity of a lengthy shutdown period of the reactor. Sometimes, a dual fiber optic system is not required. For instance, if a laser is focussed onto a suspicious-looking rock at the roadside, the ablated material may be excited and the light emitted used to identify the components. Quantification may not be easy and light collection may have poor efficiency, leading to low sensitivity. However, such a system will be capable of identifying whether the suspicious-looking rock is indeed a rock, *i.e.* it contains large amounts of Al, Si, Fe, Ca, Na, *etc.*, or whether it contains large amounts of C, H and N, *i.e.* it may well be an explosive device. Examples have even been published of LIBS being used to analyse materials on the seabed. This could potentially save dive/recovery teams significant amounts of time and money by identifying the materials' constituents before they are raised to the surface. When fiber optics are used, it should be noted that there is a loss of laser intensity which is proportional to the length of the fiber. Therefore, the excitation potential will decrease the further away the sample is from the laser source. This though is a small problem in comparison with the advantages obtained. A review of underwater

LIBS analyses of solid materials was presented recently by Matsumoto and Sakka.[19] The review contained 154 references and discussed the fundamental theory, problems and practical limitations of such analyses. The spectral lines can become distorted because of the confinement of the plasma formed. The methods developed to overcome this problem were also reviewed. It is therefore possible to use LIBS in both a qualitative and quantitative manner.

The on-line/at production line analysis of materials is also a big advantage. If a manufacturing process can use LIBS while the materials are being made, then time and money will be saved when compared with having to take a sub-sample to a laboratory for analysis. This is especially true for industries such as the steel industry, where previously an analyst may have to collect some of the molten material (an unpleasant task in itself) and then take it to a laboratory for analysis. If the LIBS analysis identifies that the steel alloy components are not in the correct ratio, this can be corrected almost immediately rather than having to wait for the results from other analytical techniques. Similarly, the recycling industry may use LIBS for rapid identification purposes. This may be in the scrap metal industry or even the identification of different polymers (through the C, H and potentially N and O ratios). A recent review (475 references) of the in-line analysis of industrial materials was presented by Pedarnig et al.[20]

The technique is capable of analysing solids, liquids and gases. It should be noted though that the analysis of liquids or of solids submerged in a liquid can lead to problems. For instance, the liquid may quench much of the energy input to excite the sample leading to a decrease in sensitivity. A recent overview, citing 128 references, of in situ on-line LIBS analysis in an environmental atmosphere was presented by Zhang and Liu.[21] In this overview, the determination of atmospheric particulate matter, halides in volatile organic compounds and atmospheric S in the environment was reviewed. Problems associated with these analyses were also highlighted. Another review summarised the use of LIBS in gas detection.[22]

The technique is generally regarded as being minimally destructive of the sample. This is because it produces a crater of diameter 30–400 μm (depending on the material type, laser energy, how well the laser beam interacts with the sample, etc.). This means that the damage caused may not even be visible to the naked eye. Since such a small area is analysed per laser shot, it also enables spatial analysis of surfaces. This can be useful when mapping is required or when different materials have been used on the same sample, e.g. different coloured paints on a painting or a ceramic.

A further advantage LIBS has is that it is very sensitive to low atomic number elements, such as Li, Be, *etc.* This compares very favourably with XRF, which exhibits poor sensitivity to lighter elements and, in some cases, cannot detect them at all.

Being capable of providing both quantitative and qualitative data is another advantage. In many of the examples given above, the exact ratio of elements in a sample may not be required. For instance, is the material found at a shipwreck gold or bronze? Simple qualitative analysis could provide the answer very rapidly.

7.2.2 Disadvantages

There are several disadvantages associated with LIBS. In some ways, it is still developing and so is experiencing many of the same problems that portable XRF experienced some years ago. The main problem is that of calibration. This is especially true for the standoff or at-line analyses. One of the main problems with the calibration, especially for quantitative analysis, is that the amount of sample excited will vary depending on the sample type. For instance, consider the analysis of plant material and ceramic. The ceramic material is harder, denser and has a higher melting/boiling point. Therefore, less material will be excited meaning that even if the two sample types produce the same intensity of emitted light for a particular element, that would have emanated from a different amount of sample. Analytical sensitivity is affected by different temperatures and pressures, different distances between the sample surface and the collection optics, *etc.* Calibration is not so problematic for the commercial benchtop instruments, although a calibration for each sample type is required for optimal performance. Similarly, the portable instruments require several different calibrations, depending on the application. Numerous calibration models have been developed which have attempted to address this problem. Several recent reviews have also covered this topic. Included in this number is one by Wang *et al.* who overviewed the recent developments in chemometric calibration methods for modern spectroscopy.[23] Although the focus of this review was not entirely LIBS (infra-red, UV-visible spectroscopy, nuclear magnetic resonance and others were also included), the methods described, *e.g.* multivariate methods, artificial neural networks, *etc.*, are very applicable. Another review was presented by Costa *et al.* who used 93 references to discuss the use of methods such as matrix-matching calibration, internal standardization, standard addition, multi-energy calibration, one-point gravimetric standard addition, one-point and multi-line calibration, slope ratio calibration, two-point calibration transfer, single-sample calibration, multiple linear regression, principal component regression, partial least squares and artificial neural networks.[24] Many of the methods use a variety of certified reference materials; some of which are used to obtain a sort of concentration–response algorithm. This is often called "training" the method. Other reference materials are then used to test the algorithm to assess its accuracy prior to using it on real "unknown" samples. Two recent reviews have also discussed the advances in calibration-free LIBS analyses.[25,26] The latter publication covered several topics and gave a brief description of the basic theory of calibration free LIBS. Several modified methods and variants proposed to overcome the non-stoichiometric ablation, self-absorption effects and high algorithmic complexity were also discussed. The method does not require standards to quantify the concentration of analytes in different sample types. Instead, it relies on mathematical algorithms including the Boltzmann plot method, the Saha–Boltzmann plot method and the column density Saha–Boltzmann plot method. These more theoretical methods usually require measurements of electron number density and other fundamental measurements prior to the analysis of real samples. Intense research has been undertaken over the last 4 or 5 years into the problems associated with calibration and decent advances have been made. The problems are exacerbated because the signals are sample dependent (*i.e.* one calibration may well not suit all materials) and can also vary according to temperature and pressure.

As discussed above, the sensitivity may be adversely affected by non-optimised operating conditions. For standoff LIBS, the sensitivity is often decreased further because of laser intensity losses along fiber optics and even poorer light collection efficiency. As a result, the sensitivity is usually lower than that of ICP–OES. The sensitivity of halogens is particularly poor. It is possible that with time, sensitivity may improve significantly – especially if more powerful/intense lasers are used. This may have the adverse effect of increasing costs though. For benchtop instruments, detection limits usually lie in the range of 1–100 mg kg^{-1}. These are nearly an order of magnitude worse compared with those obtained using XRF instruments but, as discussed previously, it does have better sensitivity for light elements. Most ICP–OES applications require a sample dissolution to be undertaken prior to analysis. This results in a significant dilution factor, *e.g.* 1 g of material being acid digested and then diluted to 100 mL results in a ×100 dilution. Under such circumstances, the limits of detection for LIBS and ICP–OES may not be very different (although this is element specific).

Another problem associated with LIBS is that it is adversely affected by rough or uneven surfaces. If the surface is rough, even at the semi-microscopic scale, then the laser may irradiate both the surface and a slightly lower level. The light collection efficiency would therefore vary slightly with depth and accounting for this can be problematic. Similarly, a curved surface may also cause problems because the plasma formed will not be uniformly equidistant from the collection optics. A recent paper by Yang *et al.* discussed the effects of sample surface morphology on LIBS signals.[27]

Another disadvantage of LIBS is that it uses lasers, *i.e.* ionising radiation. Users should be careful to ensure that safety protocols are adhered to at all times to prevent injury. In a similar manner to portable XRF instruments, care should be taken with portable LIBS instruments to ensure that legs, hands and eyes are not accidentally laser ablated. It should be emphasised that, as with portable XRF instruments, portable LIBS instruments are perfectly safe when used correctly.

One other drawback of LIBS is that it can suffer from severe self-absorption effects for analytes that are present at high concentrations. This is a further complicating factor for calibration and strategies to help compensate for it are a significant source of research.

The sample size is not a limiting factor for portable instruments. However, benchtop instruments will have a sample compartment and a lid that may limit the size of the sample to be analysed. Some instruments are modular. This means that different sample holders or spectrometers may be "attached" to the laser system. If a very large module is used, this could potentially enable larger samples to be analysed.

7.3 Hints and Tips

The operating conditions require optimisation – especially for instruments made in-house. This may mean optimising parameters such as the inter-pulse duration (also known as the "gate time") for double pulse LIBS, the frequency of the laser used (in general, picosecond pulses are preferred to nanosecond), light collection and transmission optics need to be aligned, *etc.* All of the current hand-held instruments use a Nd:YAG laser usually operating at the primary resonance wavelength (1064 nm), but the instruments made in-house can be more exotic. Although signals can be obtained with little optimisation, the best limits of detection and sensitivity require care and attention.

In general, double pulse LIBS analyses offer better sensitivity than single pulse versions. This is because the first laser pulse vaporizes the sample and the second laser pulse, that is fired several microseconds later, excites the already partially excited cloud of material. However, when this approach is adopted, careful optimisation of the time between the two laser pulses is required to obtain optimal detection limits, sensitivity and precision. It is important to note that this time may well not be the same for every sample type and will vary according to the sample's physical properties such as its density, hardness, boiling point, *etc.* Another application that benefits from the use of double pulse LIBS would be underwater analysis. The water quenches the signal from the first pulse at a rate of 10 times faster than that of air. The double pulse technique helps improve sensitivity significantly. Double pulse LIBS is a relatively new development and research work is still ongoing in this area.

The technique analyses only one tiny spot at any one instance and therefore relies on complete homogeneity of the sample for an accurate overall result. Since this is frequently not the case, many analyses of the same sample are usually undertaken using numerous laser firings over an area and then the data are averaged. Often, between 300 and 500 spectra are collected over a period of 3–5 minutes. This is one of the explanations of the relatively poor precision that is obtained when compared with techniques such as XRF or ICP–OES that are better suited to bulk analysis.

Since LIBS is essentially a surface analysis technique, it is important that the surface be free from extraneous material, *e.g.* soil on an archaeological sample or rust on a steel sample. Surfaces should therefore be cleaned so that a reasonable area that is representative of the sample bulk can be analysed. In some cases, *e.g.* aluminium alloys, surface oxide layers should be removed immediately prior to the analysis to ensure that "the bulk" is analysed representatively. Some instrument types use a very high-powered laser that effectively burns through the contaminated layers automatically before the measurement commences. Under these circumstances, a separate cleansing of the material is not required.

As mentioned previously, numerous calibration models have been tested. In general, multivariate models yield more accurate data and with higher precision than univariate models. Chemometric techniques such as LASSO or partial Least Squares Discriminant Analysis are often employed to aid multivariate calibration. Other techniques, *e.g.* Principal Component Analysis, *etc.* can be used for material classification. This is particularly useful for differentiating between samples of similar composition, *e.g.* different steel types.

Numerous benchtop models are available. Most of these come with pre-calibrations stored, others also enable calibration to be made in-house and to be customised for specific sample types. Some instrument types enable depth-profiling analyses. Others may be customised so that they can operate in harsh or unpleasant atmospheres. It is important for laboratories about to purchase an instrument that they discuss with the manufacturers what their requirements are so that they can ensure that specifications can be met. Some benchtop instruments have dual use, *i.e.* as well as a LIBS analyser, they also contain a unit capable of Raman spectrometry. This is of particular use for forensic materials or the analysis of archaeological materials. The LIBS and Raman analyses may be undertaken sequentially or simultaneously. Other instruments have been developed that offer one specialised function, *e.g.* the sorting of non-ferrous alloys.

A relatively new development that has the potential to improve sensitivity and lower detection limits significantly is the use of nanoparticle enhanced LIBS (NELIBS). The metal-based nanoparticles help improve the laser beam interaction with the sample hence increasing sensitivity. A first review into nanoparticle enhanced LIBS was presented by Dell'Aglio *et al.*[28] The review contained 51 references and focussed on the mechanisms involved, sample preparation methods, nanoparticle concentrations required and some of the early applications. The enhancement is very variable and, depending on the sample type and nanoparticles used, can be anywhere between only four or five up to over 100-fold.

References

1. Z. Wang, M. S. Afgan, W. L. Gu, Y. Z. Song, Y. Wang, Z. Y. Hou, W. R. Song and Z. Li, *TrAC, Trends Anal. Chem.*, 2021, **143**, 116385.
2. L. B. Guo, D. Zhang, L. X. Sun, S. C. Yao, L. Zhang, Z. Z. Wang, Q. Q. Wang, H. B. Ding, Y. Lu, Z. Y. Hou and Z. Wang, *Front. Phys.*, 2021, **16**, 22500.
3. G. S. Senesi, R. S. Harmon and R. R. Hark, *Spectrochim. Acta, Part B*, 2021, **175**, 106013.
4. eBook-LIBS-Technology-for-Non-Scientists.pdf (nitonuk.co.uk).
5. D. A. Goncalves, G. S. Senesi and G. Nicolodelli, *Trends Environ. Anal. Chem.*, 2021, **30**, e00121.
6. A. Botto, B. Campanella, S. Legnaioli, M. Lezzerini, G. Lorenzetti, S. Pagnotta, F. Poggialini and V. Palleschi, *J. Anal. At. Spectrom.*, 2019, **34**, 81–103.
7. G. S. Senesi, J. Cabral, C. R. Menegatti, B. Marangoni and G. Nicolodelli, *TrAC, Trends Anal. Chem.*, 2019, **118**, 453–469.

8. P. Yang, G. R. Fu, J. Wang, Z. Y. Luo and M. Y. Yao, *J. Anal. At. Spectrom.*, 2022, **37**, 1948–1960.
9. J. L. Gottfried, F. C. De Lucia, C. A. Munson and A. W. Miziolek, *Anal. Bioanal. Chem.*, 2009, **395**, 283–300.
10. L. M. Narlagiri, M. S. S. Bharati, R. Beeram and D. Banerjee, *TrAC, Trends Anal. Chem.*, 2022, **153**, 116645.
11. Q. Q. Wang, W. T. Xiangli, G. Teng, X. T. Cui and K. Wei, *Appl. Spectrosc. Rev.*, 2020, **56**, 221–241.
12. S. Legnaioli, B. Campanella, F. Poggialini, S. Pagnotta, M. A. Harith, Z. A. Abdel-Salam and V. Palleschi, *Anal. Methods*, 2020, **12**, 1014–1029.
13. O. A. Al-Najjar, Y. S. Wudil, U. F. Ahmad, O. S. Baghabra Al-Amoudi, M. A. Al-Osta and M. A. Gondal, *Appl. Spectrosc. Rev.*, 2022, **58**, 687–723.
14. K. Liu, C. He, C. W. Zhu, J. Chen, K. P. Zhan and X. Y. Li, *TrAC, Trends Anal. Chem.*, 2021, **143**, 116357.
15. Q. Zeng, J. B. Sirven, J. C. P. Gabriel, C. Y. Tay and J. M. Lee, *TrAC, Trends Anal. Chem.*, 2021, **140**, 116280.
16. R. A. Crocombe, *Appl. Spectrosc.*, 2018, **72**, 1701–1751.
17. *Laser Induced Breakdown Spectroscopy (LIBS): Concepts, Instrumentation, Data Analysis and Applications*, ed. V. K. Singh, D. K. Tripathi, Y. Deguchi and Z. Wang, 2023, ISBN: 9781119758402.
18. A. Limbeck, L. Brunnbauer, H. Lohninger, P. Porizka, P. Modlitbova, J. Kaiser, P. Janovszky, A. Keri and G. Galbacs, *Anal. Chim. Acta*, 2021, **1147**, 72–98.
19. A. Matsumoto and T. Sakka, *Anal. Sci.*, 2021, **37**, 1061–1072.
20. J. D. Pedarnig, S. Trautner, S. Grunberger, N. Giannakaris, S. Eschlbock-Fuchs and J. Hofstadler, *Appl. Sci.*, 2021, **11**, 9274.
21. Q. H. Zhang and Y. Z. Liu, *At. Spectrosc.*, 2022, **43**, 174–185.
22. Y. X. He, W. Q. Zhou, C. Ke, T. Xu and Y. Zhao, *Spectrosc. Spectral Anal.*, 2021, **41**, 2681–2687.
23. H. P. Wang, P. Chen, J. W. Dai, D. Liu, J. Y. Li, Y. P. Xu and X. L. Chu, *TrAC, Trends Anal. Chem.*, 2022, **153**, 116648.
24. V. C. Costa, D. V. Babos, J. P. Castro, D. F. Andrade, R. R. Gamela, R. C. Machado, M. A. Speranca, A. S. Araujo, J. A. Garcia and E. R. Pereira, *J. Braz. Chem. Soc.*, 2020, **31**, 2439–2451.
25. N. Zhang, T. X. Ou, M. Wang, Z. J. Lin, C. Lv, Y. Z. Qin, J. M. Li, H. Yang, N. Zhao and Q. M. Zhang, *Front. Phys.*, 2022, **10**, 887171.
26. Z. L. Hu, D. Zhang, W. L. Wang, F. Chen, Y. B. Xu, J. F. Nie, Y. W. Chu and L. B. Guo, *TrAC, Trends Anal. Chem.*, 2022, **152**, 116618.
27. L. Yang, Y. H. Zhang, Z. Zhang, Y. C. Li, Y. Xiang, J. T. Dong, Y. Q. Wei, S. T. Chang and R. S. Lu, *J. Anal. At. Spectrom.*, 2022, **37**, 1642–1651.
28. M. Dell'Aglio, R. Alrifai and A. Giacomo, *Spectrochim. Acta, Part B*, 2018, **148**, 105–112.

8 Vapour Generation and Atomic Fluorescence Spectrometry

8.1 Introduction

Since quantification using vapour generation as a means of sample introduction also requires the preparation of a calibration curve, readers should also consult the potential errors in the making of standards section at the beginning of the chapter devoted to AAS (Chapter 4).

The general procedure is to mix the liquid sample (or digested material if it had started as a solid) with a chemical reagent and then purge the vapour from the mixture using a flow of argon. The vapour is then transported to the atomic spectrometric detector leaving the remaining liquid to be transported to waste. The large majority of vapour generation applications use a chemical means of vapour generation, typically using sodium tetrahydroborate. The tetrahydroborate reduces/changes analytes to one of their reduced "hydride" forms but may also transform some analytes to their elemental state. The equipment used and the principles are similar for both. Hydride generation is therefore just a form of vapour generation. The fundamentals of the technique are outside the scope of this text, but may readily be found elsewhere, *e.g.* in the extensive textbook by Dedina and Tsalev.[1] Although elderly and concentrating very much on atomic absorption detection, the book is full of useful fundamental information,

Practical and Technical Guides for Laboratory-based Chemists No. 1
Atomic Spectrometric Methods of Analysis
By Andrew Fisher
© Andrew Fisher 2025
Published by the Royal Society of Chemistry, www.rsc.org

theory and early applications and would be an excellent starting point for someone new to the area. A much more recent book, edited by D'Ulivo and Sturgeon, is also an excellent source of information.[2] It gives updated reaction mechanisms (useful since the ones first postulated were flawed) and discusses assorted methods of chemical vapour generation, vapour generation in non-aqueous solvents, interferences, *etc.* There are alternatives to vapour generation using tetrahydroborate. Such alternatives were reviewed in 2010 by Wu *et al.*[3] This review, with 155 references, focussed on photochemical methods, halide generation, Grignard reagents for alkylation, boranes and the generation of volatile chelates. Although the majority of vapour generation techniques use chemical processes, some use electrochemical methodology. This was reviewed by Laborda *et al.* in 2007.[4]

8.2 Advantages and Disadvantages of Vapour Generation as a Means of Sample Introduction

The advantages and disadvantages of vapour generation as a means of sample introduction are summarised in Table 8.1 and are discussed in more detail in the text below.

8.2.1 Advantages of Vapour Generation

Vapour generation may be coupled with most atomic spectrometric detection techniques. Historically, it was perhaps used most with flame AAS systems in an attempt to improve sensitivity. This is not necessarily the case nowadays, with vapour generation routinely being

Table 8.1 Advantages and disadvantages of vapour generation as a means of sample introduction.

Advantages	Disadvantages
Separates analyte from a matrix that may cause spectral interferences thereby improving selectivity	Not applicable to all analytes
May be coupled with most atomic spectrometric detectors	Not applicable to all chemical species of an analyte
Increases sensitivity and improves LOD significantly	Open to potential interferences from transition and precious group metals
May be used for some analytes to obtain crude speciation/toxicity data	Can use significant volumes of high purity reagents

coupled with ICP–OES and ICP–MS instruments. It has even been coupled with electrothermal AAS instruments, but this is far from routine.

One of the most important advantages of vapour generation is that it offers greatly improved sensitivity and lower limits of detection than conventional nebulisation of liquid samples. Since the analytes are turned into a gas, their transport to the atom cell is relatively easy and can be achieved using a flow of inert carrier gas, *e.g.* argon, with an efficiency of close to 100% (assuming all precautions have been taken to transform all chemical forms of the analyte into a volatile form). This compares very favourably with that of a conventional nebuliser/ spray chamber assembly that has a transport efficiency of aqueous-based samples of 10–15% for a flame AAS system and as little as 1–2% for an ICP–OES or –MS instrument. Since more analyte reaches the atom cell per unit time when in a vapour or gaseous state, improved detection limits result. If a quartz T-piece is installed on a flame AAS burner head, sensitivity and detection limits may be improved further. A typical setup for this is shown in Figure 8.1. The analyte is swept from the vapour generation unit using the inert argon carrier gas and enters the T-piece through the stem. The flame heating the head of the T-piece then dissociates the analyte vapour (if it is in molecular form) and the analyte atoms can then absorb the light from the hollow cathode lamp which passes through the T-piece. Under normal circumstances and with conventional flame AAS analyses, the analyte spends only a few milliseconds in the light beam. However, when atomization occurs in a quartz T-piece, the analyte atoms spend much longer in the light beam because they have to escape from the tube from either end. Since the residency time is greatly increased, the amount of absorption increases. A further sensitivity improvement of between 5- and 10-fold is therefore usually obtained.

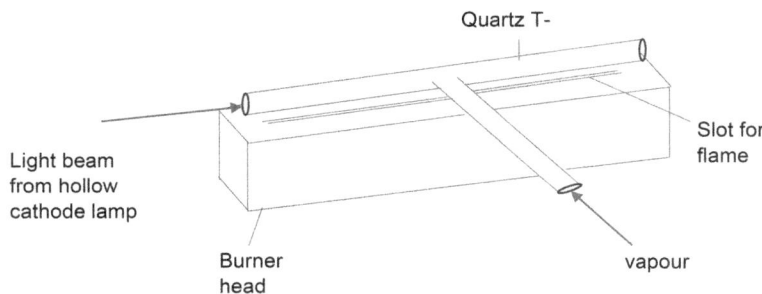

Figure 8.1 A schematic of the T-piece on a burner head.

A similar arrangement using a quartz-slotted tube atom trap also leads to improved sensitivity for flame AAS determinations. Using vapour generation can therefore lead to sensitivity improvements of 50–100 fold with a corresponding decrease in limits of detection. This enables analytes such as As or Se to be determined using a relatively inexpensive instrumental setup such as flame AAS with detection limits being obtained closer to those obtained using a far more expensive system such as ICP–MS (with conventional sample introduction).

Another big advantage is that the analytes are separated from the sample matrix during the vapour formation stage. This enables a determination that is relatively free from spectral interferences. The geometry and design of gas–liquid separators vary according to the manufacturer, with some even being made in-house. It is also possible to use the spray chamber of a flame AAS or ICP–OES/–MS instrument as the gas–liquid separator. Whatever the design, the function is the same. Even very complex sample matrices such as seawater or industrial brines may be analysed directly since the analyte(s) are largely separated from the matrix and the concomitants that could potentially cause spectral interferences. The process is very quick and is therefore a significant improvement over the use of many of the other methods of analyte preconcentration/matrix removal such as the online use of flow injection chelating columns/switching valves that are often required for many analytes that do not form hydrides. Some of the other methods used to separate analytes from sample matrix were reviewed recently by Yang et al.[5] The use of gas–liquid separators and a few potentially useful hints and tips will be discussed later in this chapter.

One of the disadvantages of vapour generation is that not all species of an analyte form a vapour, and therefore, if "total" concentrations are required, significant sample preparation methods may be necessary to convert all of the analyte into the same vapour forming species. However, under certain circumstances, this disadvantage can be used as an advantage. For example, for As, the most toxic species (arsenite and arsenate) as well as the lesser toxic ones such as monomethylarsonic acid and dimethylarsinic acid all form hydrides (albeit with different kinetics and efficiencies). However, the non-toxic species such as arsenobetaine and many of the arsenosugars do not form hydrides. A crude toxicology assessment may therefore be made by determining "total" As using conventional sample introduction into ICP–MS (for instance) followed by an assessment of the toxic proportion using vapour generation into ICP–MS. No further sample treatment is required meaning that the analyses are relatively rapid.

Although no species information will be obtained, the analyst can assess a worst case scenario by assuming that all of the reducible species are present as arsenite (the most toxic form). Hence, a relatively rapid assessment of the toxicity of a foodstuff may be made without recourse to a full speciation analysis, with all of the problems associated with it (different sample preparation methods, purchasing different analyte species for standardisation, analyte species separation methodologies, *etc.*). This and other "screening" methods of analysis were discussed by Foulkes in a book chapter in 2003.[6]

Vapour generation may also be used in combination with cryogenic trapping to obtain a preconcentration and to produce speciation data. Trapping the vapours produced from a sample in a U-tube immersed in liquid nitrogen and then removing it and allowing it to warm to room temperature enables the different species' vapours to evaporate at their own boiling point. A crude gas chromatographic system is therefore created prior to an atomic spectrometric detection method. This method can be rapid but requires standards of individual species to be purchased for quantification (and identification) purposes. In common with other vapour-forming methodologies, it is of no use when attempting to determine analytes or some species of analytes that do not form vapours.

8.2.2 Disadvantages of Vapour Generation

Unfortunately, vapour generation methods do have some disadvantages. For routine determinations, vapour generation has historically been confined to metalloids (*e.g.* As, Ge, Sb, Se and Te) and to the metals Hg and Pb. However, it has been extended to many transition metals in some research papers. The methodology required for individual analytes varies slightly, with many requiring the use of a tetrahydroborate (also known as borohydride) solution stabilised in a sodium hydroxide medium and a high purity acid stream as well as the sample in a liquid state. The concentration of the acid required varies according to the analyte, so if more than one analyte is to be determined simultaneously, then compromise conditions may be required. By default, compromise conditions may not be optimal for any of the analytes and hence the best sensitivity and LOD may not be obtained. However, the time saving obtained with simultaneous analysis partially compensates for this and an improvement in sensitivity when compared with standard nebulisation will still be obtained. For those applications where the determination of other, non-vapour-forming, analytes is also required then it is possible to use a multi-elemental technique, *e.g.* ICP–MS, ICP–OES or some flame AAS instruments where the vapour is introduced directly to the atom/ion cell, but other analytes and an internal standard may be introduced through a conventional nebuliser/spray chamber assembly that is attached *via* a Y- or T-piece (see Figure 8.2). For the determination of Hg, stannous chloride is often preferred for the reduction process rather than tetrahydroborate. Preparation of a suitable concentration of stannous chloride solution can take time because of its reluctance to dissolve which can delay the analysis. Rather than forming hydrides, *e.g.* AsH_3 or SeH_2, the reducing properties of stannous chloride allow the formation of elemental Hg vapour. In general, tetrahydroborates are preferred reducing agents because they act more rapidly. Another drawback of stannous chloride is that it is easily oxidised by air, meaning that it is not stable. Detection can be achieved using the majority of atomic spectrometric methods. It should be noted though that Hg can also form a vapour through the use of tetrahydroborate and so if numerous analytes require determination, this could be the way forward for a more rapid simultaneous analysis.

One of the biggest disadvantages is that not all of the chemical forms of an analyte form a vapour. An example is for As, where both the inorganic forms (arsenite and arsenate) form hydrides (albeit with different kinetics), whereas the species commonly experienced in fish samples *e.g.* arsenobetaine or in seaweeds, *e.g.* arsenosugars, do not.

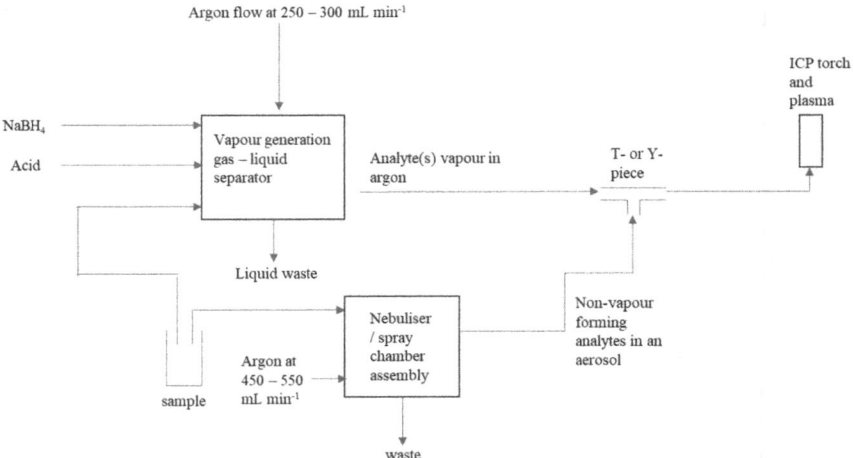

Figure 8.2 A schematic of vapour generation coupled with normal nebulisation into ICP-based instruments.

This means that a simple acid digestion followed by vapour generation when attempting to measure total As concentration is a very quick way to obtain incorrect data. With As, there is also a significant number of other species that form a hydride (*e.g.* monomethylarsonic acid and dimethylarsinic acid). However, the hydrides form with different efficiency and so will not produce the same signal intensity as an inorganic calibrant, hence potentially decreasing accuracy of the measurement. Selenium also has its issues, with selenite (Se^{IV}) forming a vapour whereas selenate (Se^{VI}) and selenoamino acids such as selenomethionine *etc*, not forming hydrides. Mercury may also present problems with the methylated species forming a volatile species more slowly than inorganic forms. This means that when "total" concentrations of the analytes are required, further sample manipulation steps may be required. This may involve the use of a far more powerful oxidising agent than nitric acid to dissolve the samples in the first instance, *e.g.* perchloric acid. Alternatively, once the sample has been digested using the relatively safe mixture of nitric acid and hydrogen peroxide, a further treatment using a persulfate may be required to destroy the organic species. Another alternative for the refractory As species is to use photolysis. The use of these other steps obviously increases sample preparation time, potentially introduces contamination or, if the sample is digested using perchloric acid, has safety implications and requires specialist facilities. The efficiency of each step should be studied so that the minimum time and/or concentration of reagents required for complete analyte conversion to an inorganic form is established. Once the analytes have been brought into ionic form they should all then be converted to the chemical form that is most efficient at vapour generation. This may involve the reduction of As^V to As^{III}, *e.g.* using L-cysteine or potassium iodide or by reducing Se^{VI} to Se^{IV} by boiling with strong (3–6 M) hydrochloric acid.

High concentrations of concomitants do not cause spectral interference during the detection stage. Instead, the transition metals mop up much of the reducing potential during the vapour generation process, leaving insufficient to convert the analytes into vapour form completely, thereby causing a chemical interference. Matrices such as ferrous and non-ferrous metals and alloys are obviously particularly problematic, although other materials, *e.g.* blood, soils and sediments that contain high levels of iron can also be problematic. The problem may be overcome through the addition of reducing agents such as ascorbic acid or complexing and masking agents such as 1,10-phenanthroline, thiocyanate, ethylenediaminetetraacetic acid (EDTA), diethylenetriaminepentaacetic acid (DTPA), 8-hydroxyquinoline, tartrate or thiourea. It should be noted though that thiourea should not be used if Hg is an analyte because a stable compound will be formed preventing Hg from forming a vapour. These complexing agents bind the transition metals hence stabilising them and preventing their reduction, leaving the full reduction potential of the tetrahydroborate to reduce the analyte(s). The precious group metals interfere in a slightly different manner. Here, stable intermetallic compounds, *e.g.* Pd–Se may form. This again prevents the analyte from producing a vapour.

The most frequently used reagent that is used in high volume is hydrochloric acid. A good quality reagent can be expensive and so when large volumes of it are used, the overall analysis becomes expensive. During hydride generation, hydrochloric acid is used along with sodium tetrahydroborate to produce the volatile analytes. *N.B.* the term "reduced to" should not really be used because vapours such as plumbane (PbH_4) have Pb in a higher oxidation state than Pb^{2+}, or As III compounds form AsH_3 where no change of oxidation state is involved. The concentration of hydrochloric acid required varies depending on the analyte but can be up to five or six molar, *i.e.* it is nearly a two-fold dilution of the commercially available concentrated acid that you start with. This can be used at a rate of several mL per minute. Therefore, a full day's work can consume a full 2.5 L bottle of the required higher quality reagent which may be prohibitively expensive for some laboratories. A potential cost-saving measure could be to use sub-boiling distillation to clean up a lower grade of acid.

8.3 Useful Hints and Tips for Vapour Generation

As stressed in previous sections, if "total" analyte concentrations are required to be determined then, depending on the sample type, extensive sample manipulation may be required to convert all of the analyte species into the same reducible form. There are numerous ways of achieving this including the use of chemical reductants, *e.g.* ascorbic acid, potassium iodide or L-cysteine; all of which have been used to reduce As^V to As^{III}. The use of very powerful oxidising agents has also been used to destroy organic compounds of analytes, *e.g.* perchloric acid or alkaline persulfate. The use of very powerful oxidising agents such as these are required to destroy stable compounds such as arsenobetaine, many arsenosugars and seleno-amino acids such as selenomethionine and selenocysteine. It should be stressed that a typical nitric acid digestion of a sample (even with hydrogen peroxide) will have minimal effect on some refractory compounds such as arsenobetaine. A nitric/peroxide digestion is therefore excellent at destroying fish tissue (beware the formation of nitrated glycerides that may be explosive) and will be completely adequate for sample analysis using conventional nebulization ICP–MS, ICP–OES or AAS. On its own though, it is inadequate for determining "total" As using vapour generation. The use of extremely powerful oxidants such as perchloric acid is likely to destroy the organic compound leaving the As or Se in the As^{V} and Se^{VI} state. These therefore require pre-reduction prior to hydride formation. For As, the ascorbic acid, L-cysteine or potassium iodide is adequate. Se^{VI} cannot form a hydride and so a further reduction (typically achieved by boiling the sample with 3–6 M HCl) is required to transform it into the Se^{IV} state. The use of perchloric acid can be hazardous and therefore many laboratories try to avoid it. The use of a mixture of hydrobromic acid and bromate has also been shown to destroy numerous organic selenium compounds. This has the added advantage of leaving Se in the Se^{IV} state. Other workers have used photolysis to destroy the organic compounds, *e.g.* the work by Rubio *et al.*[7] The kinetics of the chemical treatments may be accelerated through the use of microwave irradiation. Examples include the destruction of As compounds following their separation using a chromatographic column[8] and the on-line reduction of Se^{VI} to Se^{IV} prior to vapour generation and AFS detection.[9]

It should be noted that just because a set of vapour generation conditions work with one type of detector, they may be far from ideal for another. When coupled with AAS, an excess of hydrogen is not usually problematic since it burns harmlessly at the open ends of the quartz tube. For AFS detection, it may even be helpful in that it may be used to form the entrained hydrogen flame used as the atom cell. However, a large excess of hydrogen may be problematic for a plasma-based instrument, where it could change the thermodynamic properties of the plasma and may lead to instability. **The generation conditions may therefore have to be optimised with the detector in mind**.

The use of argon as a sweep gas to transport vapour(s) to the atom cell should also have its flow rate optimised. Typically, a flow of 250–300 mL min^{-1} is used. This is insufficient for maximum sensitivity for most plasma-based instruments and may even be insufficient to "punch" the plasma, *i.e.* an annulus may not be formed. It is often necessary to use a "Y-piece" with a make-up argon flow of $\sim$500 mL min^{-1} from the nebuliser/spray chamber assembly to give a total of 0.75–0.85 L min^{-1}. This has the added advantage of enabling an internal standard to be nebulised, which may help improve signal stability. Similarly, it would also allow non-vapour generating elements to be determined simultaneously with the vapour forming ones.

Excess frothing in the gas–liquid separator can be problematic as it can cause signal instability. There are several options available to minimise this. A few drops of a higher alcohol, *e.g.* octanol or decanol may be placed in the gas–liquid separator. This helps calm the frothing. Similarly, glass beads placed in the U-tube of the separator may have the same effect. If both of these fail to calm the frothing, then perhaps a lower concentration of the tetrahydroborate should be used.

Although the use of a gas–liquid separator decreases the amount of liquid entering the atom cell significantly, there is always a small residual amount of water vapour carried in the gas flow. This is unlikely to cause a problem for most detection systems but could severely affect signal stability during the AAS or AFS detection of some analytes, especially Hg. This is because a flame is not required to atomize the Hg vapour; being already in atomic form. Water droplets could therefore lead to an erroneously high "absorption" signal and a very noisy fluorescence signal as the light fluoresced may be diffracted by the water droplets. Such problems can be overcome by the use of a 'semi-permeable membrane drier tube' between the vapour generator and the atom cell.[10] Drier tubes have also been used for other analytes, *e.g.* As.[11] Interestingly, this paper demonstrated that the vapours of some methylated species of As were lost in the membrane drier tube. The authors therefore recommended sodium hydroxide beads in a cartridge as a better drying method. The paper is also a good example of the use of cryogenic trapping of species prior to atomic spectrometric detection.

As with all methodologies, the analyst should assess every step of the process for potential losses of the analyte, inefficiencies in the formation of vapours, potential contamination, *etc.*

8.4   Atomic Fluorescence Spectrometry

Dedicated, commercial multi-element atomic fluorescence instruments have been very limited in number. Atomic fluorescence is a technique that is used almost exclusively nowadays as a detector for vapour generation. The main components of an atomic fluorescence detector are shown in Figure 8.3. It requires a light source (that can be of higher intensity than a standard hollow cathode lamp, *e.g.* a boosted one, to aid sensitivity), an atom cell (which is usually a flame), a means of sample introduction (nowadays normally vapour generation), a method of isolating the light emitted from the atom cell, a detector and a readout system. For the instruments dedicated to act as detectors for vapour generation, the wavelength isolation device is often a filter rather than an expensive monochromator. This obviously helps keep the overall cost down. The orientation of the incident light beam and measurement of fluoresced light is normally at a right angle to ensure that light from the light source is not detected.

There are five types of atomic fluorescence. These are termed: resonance fluorescence, Stokes direct line fluorescence, stepwise line fluorescence, two-step excitation or double resonance and sensitized fluorescence. The difference between these different types is beyond the scope of this text but may be found elsewhere.[12,13] Of these different types, resonance fluorescence is the most useful analytically. This is where the excitation wavelength and the fluorescence wavelength are the same.

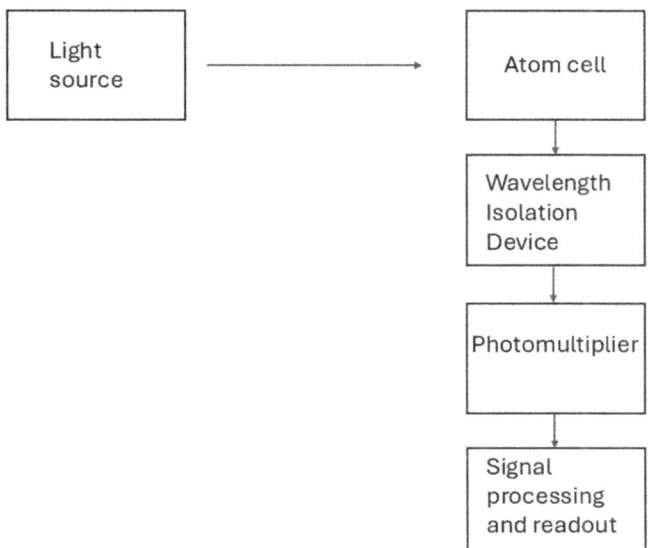

Figure 8.3 A schematic diagram of a fluorescence instrument.

8.4.1 Advantages and Disadvantages of AFS

The advantages and disadvantages of atomic fluorescence are given in Table 8.2 but are discussed in more detail in the text below.

As mentioned previously, vapour generation systems may be coupled with most atomic spectrometric detectors. This includes AAS systems (including electrothermal atomization), ICP–OES, ICP–MS and atomic fluorescence. Although atomic fluorescence is a much more selective spectrometric measurement technique in terms of analyte, when coupled with vapour generation, it becomes a (relatively) cheap extremely sensitive and reliable method of analysis. It has extremely low detection limits (at the $pg\,mL^{-1}$ level for some analytes) which is far better than emission and most absorption-based techniques. It also has an extremely long linear range which spans typically 5 or 6 orders of magnitude. This is similar to emission-based techniques and is better than absorption-based ones. There are companies that make vapour generation systems that may be coupled with other instrument types, and there are also modular instruments that may be used as fluorescence detectors of As and Se, of Hg, *etc.* Overall, these modular-based instruments are significantly cheaper than other analytical systems and require far fewer laboratory facilities, *e.g.* they do not require an acetylene fuel to form a flame. Instead, they just require a suitable means of extraction to meet any health and safety requirements. They are small and compact and may easily be transported to and then used in non-specialist laboratories, *e.g.* off-shore on an oil rig. For the determination of Hg, no flame is

Table 8.2 Advantages and disadvantages of atomic fluorescence.

Advantages	Disadvantages
Relatively simple technique compared with atomic absorption spectroscopy.	Background noise arising from radiation dispersion by flame particulate matter or water droplets.
Precision better than 1% can be achieved.	Not applicable for all compounds due to fluorescence issues.
Greater sensitivity than emission or absorption techniques.	Theoretically possible for most elements but in practise limited mainly to metalloids, as well as Cd, Hg and Pb.
Two substances that are excited at the same wavelength but emit at distinct wavelengths can easily be distinguished.	
Long linear range	

required. The Hg vapour formed simply enters the atom cell where it is excited by the high intensity light beam causing it to fluoresce. For other analytes, the hydrogen formed as a by-product of the vapour generation process may be used as the fuel of an air-hydrogen flame. A recent review of hydride generation –AFS was presented by Zou *et al.*[14] The review, which contained 142 references, focussed on the portable and miniaturized instrumentation that can be used in the field.

The main drawback of atomic fluorescence spectrometry is that it suffers from quenching and, especially for Hg, problems arising from the presence of particulates or water droplets in the atom cell. The latter arises mainly when vapour generation is used because no flame is employed to atomize the analyte.

8.4.2   Data Analysis

The procedures required for data analysis when using atomic fluorescence are the same as for the other techniques. When used as a detection method for vapour generation, comparing the experimental result for a certified material with the certified value is extremely important for many sample types. As discussed previously, many biological and some environmental samples may contain As (and other elements) in several different chemical forms; many of which do not form a vapour. If results significantly lower than those expected for the certified material are obtained, it could well indicate that one or more of the sample preparation steps is not working at optimal efficiency. If only nitric acid has been used to digest a seaweed or fish tissue, then further manipulation steps will be required to obtain the "total" As content. If photolysis is being used to split organoarsenic compounds, then a longer photolysis period may be required or a higher intensity lamp used. The same is true for Se. Another possible cause would be inefficient "reduction" to a species capable of forming a vapour. A reducing agent such as ascorbic acid could be added for As or if Se is the analyte a longer period of boiling with 6 M hydrochloric acid may be required. For other sample types, *e.g.* a steel, a ceramic, glass, or other inorganic materials, then the problem with organoarsenic or organoselenium compounds will not exist. For sample types such as these, after acid dissolution or any other sample preparation protocol, a further step such as photolysis *etc.* will not be required.

When determining vapour forming elements as well as other analytes that are introduced *via* a conventional nebuliser/spray chamber assembly to an ICP–OES or –MS instrument, care should be taken to ensure that measurement is only made when both sample streams have reached the plasma. This may involve having a slightly longer uptake time before measurement commences. Failure to wait for sufficient time is likely to lead to either the vapour forming elements or the other analytes not reaching the plasma in time. The concentration calculated for the first replicate measurement (and possibly even the second replicate if the necessary uptake time has been severely under-estimated) will be lower than for the other replicates.

The use of an internal standard should be able to help correct for slight changes in sensitivity over longer periods but will not be able to correct for errors associated with blockages of the nebuliser, *etc.* if an error is made when inserting the internal standard to all samples and standards, then this too is likely to be difficult to remedy. A simple two-fold error, *e.g.* if a 50 mL flask is used rather than a 25 mL one, then this can be remedied. However, if an air bubble is accidentally admitted, then this is impossible to correct for.

References

1. J. Dedina and D. Tsalev, *Hydride Generation Atomic Absorption Spectrometry*, John Wiley & sons, Chichester, 1995, ISBN 0471953644.
2. *Vapor Generation Techniques for Trace Element Analysis: Fundamental Aspects*, ed. A. D'Ulivo and R. Sturgeon, Elsevier, 2022, ISBN: 9780323858342.
3. P. Wu, L. A. He, C. B. Zheng, X. D. Hou and R. E. Sturgeon, *J. Anal. At. Spectrom.*, 2010, **25**, 1217–1246.
4. F. Laborda, E. Bolea and J. R. Castillo, *Anal. Bioanal. Chem.*, 2007, **388**, 743–751.
5. Q. Yang, Y. L. Zhao, H. M. Yu, X. L. Xiong and K. Huang, *Appl. Spectrosc. Rev.*, 2023, **59**, 652–677.
6. Screening Methods for Semi-quantitative Speciation Analysis; Chapter 8 (pp 591–604) by M. E. Foulkes, in *Handbook of Elemental Speciation: Techniques and Methodology*, ed. R. Cornelis, K. Heumann, J. Caruso and H. Crews, J. Wiley and Sons Ltd, Chichester, 2003, vol. 1.
7. R. Rubio, J. Alberti, A. Padro and G. Rauret, *Trends Anal. Chem.*, 1995, **14**, 274–279.
8. K. J. Lamble and S. J. Hill, *Anal. Chim. Acta*, 1996, **334**, 261–270.
9. L. Pitts, A. Fisher and S. J. Hill, *J. Anal. At. Spectrom.*, 1995, **10**, 519–520.
10. W. T. Corns, L. Ebdon, S. J. Hill and P. B. Stockwell, *Analyst*, 1992, **117**, 717–719.
11. M. Svoboda, J. Kratzer, M. Vobecky and J. Dedina, *Spectrochim. Acta, Part B*, 2015, **111**, 46–51.

12. D. J. Butcher, Atomic Fluorescence Spectrometry, in *Encyclopedia of Analytical Science*, ed. P. Worsfold, C. Poole, A. Townshend and M. Miro, Academic Press, 3rd edn, 2019, pp. 201–208, DOI: 10.1016/B978-0-12-409547-2.14531-7.
13. S. J. Hill and A. S. Fisher, Atomic Fluorescence methods and Instrumentation, in *Encyclopedia of Spectroscopy and Spectrometry*, ed. J. C. Lindon, G. E. Tranter and D. W. Koppenaal, Academic Press, 3rd edn, 2017, pp. 65–69, DOI: 10.1016/B978-0-12-803224-4.00371-X.
14. Z. R. Zou, Y. J. Deng, J. Hu, X. M. Jiang and X. D. Hou, *Anal. Chim. Acta*, 2018, **1019**, 25–37.

9   Which Technique to Use?

The choice of which analytical technique to use for an analysis is governed by numerous factors. The prime factor is which techniques are available. Not all laboratories are fortunate enough to have a full suite of techniques and may have to make do with what they have even if what they have is not the most suited to the task. Under such circumstances, the sample preparation protocol is likely to be more complex than normal, *e.g.* a lengthy and potentially dangerous acid dissolution may be required. Alternatively, a preconcentration protocol may be required if the analyte(s) are below the limit of detection of the instrumentation available.

If only one analyte is to be determined on a routine basis, *e.g.* Cd at high concentration in an electroplating waste sludge, then a simple and inexpensive flame AAS instrument would potentially be best. A simple acid dissolution of the sample followed by AAS analysis is rapid, relatively inexpensive in terms of initial outlay for the instrument and for running costs and does not require extensive training for the analyst. Investing in a much more expensive instrument capable of simultaneous multi-element determination at very low concentration would be a waste of resources unless it was envisaged that the analyte suite was to be expanded. The one potential drawback of flame AAS would be the safety aspect, with the acetylene required for sustaining the flame having to be stored outside of the building. Alternatives would include a laboratory-based or portable XRF or LIBS analyser that could analyse the sludge directly.

If a suite of analytes is to be determined, the analyst should try and identify whether all of them are likely to be at a sufficiently high

Practical and Technical Guides for Laboratory-based Chemists No. 1
Atomic Spectrometric Methods of Analysis
By Andrew Fisher
© Andrew Fisher 2025
Published by the Royal Society of Chemistry, www.rsc.org

concentration to be above the limit of detection of the instrument. The choices for analysing at higher concentrations would be XRF or LIBS (both of which are capable of analysing solid samples directly with minimal or no dilution) or ICP–OES following a suitable acid dissolution. All three techniques are capable of determining analytes at the 10 $mg\,kg^{-1}$ level. If the analytes are present at a much lower concentration, then the much more expensive technique of ICP–MS may be required. An alternative would be to undertake a pre-concentration protocol so that the analyte concentrations increase to the level at which they may be detected using ICP–OES.

Another factor to be taken into account is the ease with which the sample can be prepared for presentation to the instrument. Some sample types, *e.g.* ceramics, glasses, soils/sediments and rocks are particularly difficult to dissolve completely. If an aggressive leach is adequate, *i.e.* it is necessary only to determine the "biologically available" concentrations of analytes then either an AAS or a plasma-based instrument could be used. If "total" concentrations are required then those techniques capable of analysing solids directly or with only minimal dilution, *i.e.* XRF or LIBS may be a better option.

A further factor to consider is whether a plasma-based instrument could be used in conjunction with a solid sampling device such as a laser ablation unit or an electrothermal atomiser. These are both available commercially but at a significantly elevated cost. For instance, a good laser ablation unit would cost more than a flame AAS instrument. The laser ablation unit would though have the added advantage of minimising damage to the sample. In addition, it would offer the possibility of obtaining spatial information, *i.e.* the laser could be focussed on an extremely small spot and then moved across the surface of the sample. Such methodology would enable an elemental "map" of the surface to be constructed. A laser ablation unit would potentially also enable the analysis of extremely small targets, such as a fluid inclusion in a rock sample or inclusions in steel samples. Although an electrothermal atomiser would not yield such important information, it is still capable of analysing solid materials directly hence negating laborious or dangerous dissolutions or fusions. It would require a high precision balance so that a very accurate weight of material, often at or below 1 mg, can be introduced. This leads to one of the main drawbacks of the technique – poor precision. This can arise through poor weighing technique, *i.e.* attempting to weigh 1 mg on a four figure balance is likely to lead to significant errors since the fourth decimal place often fluctuates or wanders. Having weighed the material, it then has to be transported to the atomiser ensuring that none is lost, *e.g.* blown away by an air conditioning unit or spilled while being introduced.

If the sample is precious and needs to be preserved as much as possible, *e.g.* a forensic or heritage sample, then a minimally-damaging technique would be preferred. Examples of such techniques would be XRF (particularly the portable version), LIBS, or laser ablation followed by either ICP–OES or ICP–MS detection.

If isotopic ratio data are required, then ICP–MS would be the choice of technique, especially if the instrument available has a sector field with an array of detectors. This would be capable of simultaneous measurements and would therefore provide the most accurate ratio data. This is important for some applications, *e.g.* nuclear forensics, dating of geological samples, isotope dilution analyses, *etc.*

If water samples are to be analysed, then LIBS is unlikely to be the first choice of instrumentation to use because the plasma is quenched leading to low sensitivity. Similarly, the sample may boil and "spit" into the air above the surface partially blocking some of the light that is emitted. Many users also shy away from using XRF to analyse water samples because of the potential damage that could occur should the sample container leak. Both of these techniques are also likely to struggle to reach the required sensitivity for most analytes. A possible way to circumvent this problem would be to mix the sample with a chelating resin and a suitable buffer system which would retain the divalent ions, *e.g.* transition metals and some others. If a litre of water is mixed with 1 g of resin, stirred on a magnetic stirrer and the resin then filtered off and dried, an effective preconcentration factor of 1000 would be obtained (assuming 100% efficiency in retention). Once dried, the resins could then be analysed directly using either XRF or LIBS. Such an approach could also be used to improve limits of detection prior to FAAS, ICP–OES and ICP–MS analyses although the analyte(s) would need to be eluted from the resin in a small amount of acid so that they can be introduced to the instrument *via* the standard nebuliser/spray chamber assembly. This would result in a less impressive enrichment factor, but could still bring the analyte concentrations to levels that enable them to be determined accurately.

Portable instrumentation is of huge value if work is to be undertaken in "the field". For instance, geologists can use portable XRF or LIBS instruments to identify minerals within seconds rather than have to transport them back to a laboratory for a "proper" analysis. This therefore saves time and expense and would clearly be much more convenient.

The attributes and characteristics of each technique are summarised in Table 9.1.

Table 9.1 Attributes and characteristics of the different techniques.

Technique	Relative cost	Limits of detection	Analytical capabilities	Other comments
FAAS	Less expensive	$1-1000$ mg kg^{-1}	Requires a liquid sample. Therefore, acid dissolutions, fusions or other extractions of solid materials required. Most instruments are capable of only single element analysis at a time although some more expensive ones do have multi-element capability. Relatively short linear range (2–2.5 orders of magnitude). Can easily be automated. Requires ~1 mL of sample per analyte unless flow injection, a discrete sampler or a more expensive multi-element instrument is used. Rapid, with single analytes being determined in only a few seconds. Very few spectroscopic interferences. Other interferences are well documented and easily overcome.	Infrastructure such as fume extraction and a gas supply are required. Acetylene may have to be piped from outside the building. Most instruments require hollow cathode lamps of the analytes of interest to be purchased.
ETAAS	Moderately expensive	$0.01-1$ mg kg^{-1}	May be used for liquid or solid samples (if a modified tube is used). Short linear range (2–2.5 orders of magnitude). Slow analysis rate with each sample taking 1–2 minutes per replicate. Very low sample volume required (10–50 µL) per replicate. Few spectroscopic interferences but others, *e.g.* smoke can be	Hollow cathode lamps required for most instruments. Graphite tubes are an added expense.

Table 9.1 (*Continued*)

Technique	Relative cost	Limits of detection	Analytical capabilities	Other comments
			problematic. Optimisation of temperature program and the use of matrix modifiers help minimise these problems. Easily automated.	
ICP-OES	Moderately expensive	0.1–100 mg kg^{-1}	Unless a specialist solid material introduction device is used (laser ablation, *etc.*), it requires a liquid sample. Acid dissolutions, fusions or other extractions of solid materials required. Long linear range (5 orders of magnitude). Rapid multi-analyte (60–70)/multi-wavelength capability. Easily automated. Capable of determining 50 analytes in only 3 or 4 mL of sample. Some spectroscopic interferences but they are well documented and can be overcome by using alternative wavelengths or by judicious optimisation of background correction parameters.	Relatively expensive infrastructure *e.g.* fume extraction system, gas manifolds, air-conditioning (for best stability) and circulating water chillers all required. Also required is an argon supply. This can be delivered in cylinders (at a rental cost per week for the cylinder plus the cost of the gas) or in liquid form. The latter would require a large Dewar-style flask.
XRF	Moderately expensive (EDXRF) or expensive (WDXRF)	1–100 mg kg^{-1}	Mostly used for solid samples, especially those that are hard to dissolve, *e.g.* rocks, soils/sediments, ceramics and glasses. Analysis speed varies with technique and number of analytes. EDXRF is rapid with numerous analytes being detected	Significant infrastructure (vacuum system, gas supply and coolant water supply) required for WDXRF. Much less infrastructure required for EDXRF. Specialist sample preparation equipment also required,

Technique	Cost	Detection limit	Comments	Infrastructure
			simultaneously in 1–3 minutes whereas WDXRF is much slower and detects analytes in a semi-sequential way (possibly an hour for a suite of 25 analytes). Interferences are well documented and may be overcome or circumvented by using correction procedures or using alternative wavelengths/energies. Long linear range (6 orders of magnitude) from mg kg^{-1} to 100%.	*e.g.* a mill, a press and a fusion furnace.
Portable XRF	Less expensive	1–100 mg kg^{-1}	Used mainly for solid samples. Non-destructive technique. Therefore ideal for forensic or archaeological studies. Also ideal for screening studies *i.e.* is an analyte present or not for instance Br in plastics. Also ideal for field work. Rapid analysis potentially providing results in less than a minute. Calibrations for different applications are stored internally on the instrument. Serious errors can occur if the wrong calibration is used.	No infrastructure required, but factory re-calibration recommended every year or two. Safety aspects must be taken into account to ensure users are not exposed to X-rays.
ICP-MS	Expensive for single quadrupole instruments. Very expensive for triple quadrupole instruments	0.001–0.1 mg kg^{-1} or 0.001–1 µg L^{-1} for liquid samples.	Unless a specialist solid material introduction device is used (laser ablation, *etc.*), it requires a liquid sample. Acid dissolutions, fusions or other extractions of solid materials required. Long linear range (up to 9 orders of magnitude). Rapid multi-analyte	Relatively expensive infrastructure *e.g.* fume extraction system, gas manifolds, air-conditioning (for best stability) and circulating water chillers all required. For ultra-trace analyses the additional

Table 9.1 (*Continued*)

Technique	Relative cost	Limits of detection	Analytical capabilities	Other comments
	and extremely expensive for magnetic sectors.		(60–70)/multi-isotope capability. Easily automated. Capable of determining 50 analytes in only 3 or 4 mL of sample. Several types of interference. Most are well documented and can be overcome/circumvented by use of collision cell or reaction cell gases or by using alternative isotopes. Mathematical algorithms may also be used. Another method for interference reduction used mainly in the semiconductor industry is cool plasma ICP-MS. ICP-MS provides isotopic information.	expense of a clean room may be required. Also required is an argon supply. This was discussed in the ICP–OES section.
LIBS	Moderately expensive	mg kg^{-1} range	Used almost exclusively for solid samples. Relatively non-destructive. Rapid analysis. May provide quantitative or qualitative data. May be used in stand-off mode so is useful for hazardous samples *e.g.* explosives or radioactive materials. Specialised instruments may be used for on-line analyses. Calibration can be problematic and requires calibrants very closely matched to samples. Fairly long linear range but can be affected by self-absorption problems.	This is still a developing technique but is developing rapidly. Little infrastructure is required for standard instruments. Standoff or at-line instruments will usually require fibre-optic cables.

| Portable LIBS | Less expensive | Typically at the mg kg^{-1} level but may be significantly worse for standoff analysis. | Used mainly for solid samples. Non-destructive technique. Therefore ideal for forensic or archaeological studies. Also ideal for screening studies *i.e.* classifying different types of polymer. Also ideal for field work. Rapid analysis potentially providing results in less than a minute. Calibrations for different applications are stored internally on the instrument. Serious errors can occur if the wrong calibration is used. | The presence of a laser means that safety aspects must be considered. |

Subject Index

Page numbers in *italic* refer to figures; in **bold** to tables